Spatial Processes in Plant Communities

Czechoslovak Academy of Sciences

Spatial Processes in Plant Communities

Proceedings of the Workshop held in Liblice,
18–22 September 1989

edited by

F. KRAHULEC
Czechoslovak Academy of Sciences
Průhonice

A. D. Q. and S. AGNEW
University College of Wales
Aberystwyth

H. J. WILLEMS
University of Utrecht

SPB Academic Publishing bv, The Hague, 1990

Co-published with
Academia, Publishing House of the Czechoslovak Academy of Sciences,
Prague, Czechoslovakia

Distributors for Albania, Bulgaria, China, Cuba, Czechoslovakia, German
Democratic Republic, Hungary, Mongolia, North Korea, Poland, Rumania,
U.S.S.R., Vietnam and Yugoslavia:

Academia, Publishing House of the Czechoslovak Academy of Sciences,
Prague, Czechoslovakia

Distributors for all remaining countries:
SPB Academic Publishing bv, P.O. Box 97747, 2509 GC The Hague,
The Netherlands

CIP-DATA KONINKLIJKE BIBLIOTHEEK, DEN HAAG

Spatial

Spatial processes in plant communities : proceedings of
the Workshop held in Liblice, Czechoslovakia, 18–22
september 1989 / ed. by F. Krahulec . . . [et al.]. – The
Hague : SPB Academic Publishing. – Ill.
Ed. is co-published with Academia, Publishing House of the
Czechoslovak Academy of Sciences, Prague. – With ref.
ISBN 90-5103-041-X bound
SISO 584 UDC 581.5 NUGI 824
Subject headings: plant community patterns / ecology.

Printed in Czechoslovakia

CONTENTS

Preface

In 1985 some young plant ecologists of eastern Europe, building upon an older tradition of synecological taxonomy in vegetation study, joined together to discuss and report on the numerical aspects of their science. In that year an informal workshop was held in Budapest, Hungary, and it was followed by a similar meeting in Galanta, Czechoslovakia in 1987. Meanwhile, in 1986, the Botanical Institute at Třeboň, Czechoslovakia, had hosted a discussion on secondary, anthropogenic successions. In early 1988, ecologists at the Botanical Institute at Průhonice conceived the idea of putting together many of the topics of discussion at earlier meetings in one devoted to vegetation processes, particularly as applied to change in spatial relationships. Many contacts were made at the Vienna conference of the International Association of Vegetation Science in July 1988, and in other informal meetings during that year. The product of this planning was the very successful workshop meeting at Liblice, Czechoslovakia 18–22 September 1989, hosted and ably organized by the Botanical Institute, Czechoslovak Academy of Sciences. This volume is the record of that workshop, being the papers that resulted from the presentations and posters discussed by the participants.

That the title of that meeting was exactly the same as the title of this volume reflects the careful prediction of interest by the organizers. Every participant was able to choose his own approach to the problems of spatial processes in vegetation. There is no consensus, of course, on the value and importance of the various possible scales of research into vegetation process. Small plants or big, short distances or long: there are insights to be gained into Nature by following all these leads. Therefore we have in this volume a challenging diversity of reports, from heroic concepts of geologic and climatic influence on floristic diversity in whole mountain massifs to evaluation of miniature pattern and process in bryophytes amongst chalk grassland. The time scale can be centuries, decades, years or months. The vegetation data may be from visual evaluation, contact or biomass, and a very wide array of mathematical tools called into use in their analysis. In most papers the causality of the process of change is the focus of investigation often using experimental techniques, while in others the observation of a more static pattern is related to predicted processes. Yet these proceedings do revolve around a central subject area: that which concerns the changes in patterns of plant presence in Nature. Frequently these patterns must be considered in relation to disturbance regimes and mankind is often the main disturber. Therefore these questions are not only of academic interest; the processes can take place just as well in crops whether trees or herbs. Diseases, pollution and predators can be elements linking our study of spatial processes in the natural world with our commercial and agricultural interests.

We, the editors, want to thank the many Czechoslovak ecologists who took part in the meetings, in discussion and excursions, helping visitors to discover the delights and riches of the Bohemian landscape and vegetation. The meetings were largely chaired by participants, but we thank Prof. Emil Hadač for coming to guide one day's work, and dr. Robert Neuhäusl for the opening of the workshop. We would also acknowledge the hospitality of the Czechoslovak Academy of Sciences in whose delightful Chateau of Liblice our academic and social requirements were so amply met.

František KRAHULEC
Jo H. WILLEMS
Andrew D.Q. AGNEW
Shirley AGNEW

Introduction

František Krahulec

Institute of Botany, Czechoslovak Academy of Sciences, 252 43 Průhonice, Czechoslovakia

The inter-relationships between structure and process are amongst the most frequently studied topics in vegetation science. One of the ways of studying these relationships is to follow structural changes in through time, while a knowledge of spatial processes combines information from population and community levels so that we may explore their mutual relationships. This was the main aim of the workshop held in Liblice (CS) in September 1989 of which the proceedings are presented here.

The papers are ordered according to the increasing scale of the subject matter, and from descriptions of pattern change in time to those analysing mechanisms. Theoretical papers are included but most are case studies. The diverse material illustrates preoccupations amongst plant ecologists that we can codify as:

1. How can temporal pattern changes be described?
2. How can we combine different scales?
3. What are the underlying mechanism of temporal change of vegetation structure?
4. Can underlying mechanisms be deduced from observed patterns?

Following notes relate these topics to our collected papers in an attempt to draw together the diverse strands represented here.

It seems now that there are few problems with the description of pattern at various scales. The combination of different scales of space and/or time causes some difficulties, mainly at small scales where trends in the observations are not so clear and the appropriate mathematical techniques are not yet completely developed.

Various approaches are applied to describe temporal changes in spatial pattern. An analytical approach is used to study relatively simple relations (e.g. Hughes and Hara *et al.*) and is combined with modelling (e.g. Leemans). At a larger, landscape scale, species can be studied with respect to their behaviour in relation to the rarity of their habitats. Rabinowitz's seven forms of rarity (Rabinowitz 1981) are used by Söderström in his study on temporary habitats throughout Sweden. The use of temporal and spatial autocorrelograms seems to be an efficient method for pattern detection in individual species; however, we need similarly efficient methods at the level of the whole community.

When more complicated systems are considered some difficulties appear: for example Herben *et al.* use simple species pair replacement dynamics, both weighted and unweighted, but the results should be carefully interpreted. Another approach applying transition matrices is used by Rijnberk & During. Matus & Tóthmérész use classical plexus technique combined with ordination (PCoA) to obtain more objective results for groups of species, called coalitions or guilds. A special approach based on information theory is used by Bartha.

Several papers are devoted to grasslands (Herben *et al.*, Rijnberk & During, Rychnovská & Jakrlová) and show that, although they are usually considered stable at one scale, they consist of highly dynamic patches both in space and time. At a small scale, each plot seems to behave as a single unit with its own type of change. The system observed seems to have chaotic behaviour which produces stability at a higher spatial and temporal scales.

The analysis of causality is a difficult task. Although it is possible to generate various hypotheses, their verification needs complex further research. In some cases correlations suggest causes as with climatic conditions (Rychnovská & Jakrlová), or soil resources (Agnew & Gitay, Symonides & Wierzchowska). We must agree with Shipley and Keddy (1987) and Lepš (here) that manipulative experiments are a necessary next step in testing these hypotheses.

At higher levels of change in patterns of whole communities we can again detect various phenomena, but their explanations are based more on experience than on unequivocal evidence. It is relatively easy to follow changes in boundary zones between communities (Körmöczy & Ballogh, Mezsáros) or changes in whole communities (Daniels). Interpretation of observed changes is possible if research on underlying mechanisms is available (Willems and Bobbink).

The following points can be made to summarize these Proceedings:
1. Different scales should be considered in analyzing vegetation changes in time.
2. Each scale reflects the action of particular phenomena which need not be expected at a higher scale.
3. Causal explanations of pattern changes are difficult to provide without additional information, mainly from manipulative experiments.
4. Mathematical methods are still to be developed to describe and explain multispecies pattern changes in time.
5. Fine scale observations suggest that relatively stable systems can be the result of subsystems which are highly diverse, both in space and time.

References

Rabinowitz D. (1981): Seven forms of rarity. — In: Synge H. [ed.], The biological aspects of rare plant conservation, Chichester.

Shipley B. and Keddy P.A. (1987): The individualistic and community concepts as falsifiable hypotheses. — Vegetatio 69: 45–55.

Krahulec F., Agnew A.D.Q., Agnew S., Willems J.H. [eds.]: *Spatial processes in plant communities.* pp. 1–11.

Can underlying mechanisms be deduced from observed patterns?

Jan Lepš

Department of Biomathematics, Biological Research Centre, Branišovská 31, CS-370 05 České Budějovice, Czechoslovakia.

Keywords: Spatial pattern, Interspecific associations, Neutral model, Species-area relationship, Experiments, Hypotheses testing

Abstract. The aim of this paper is to evaluate attempts to estimate underlying mechanisms from observed patterns. The following cases are discussed:

1. The spatial pattern of a single population and interactions among individuals.
2. The spatial arrangement of two populations (statistical 'interspecific associations') and interactions between populations.
3. Analysing the distribution of population abundances and comparison with a neutral model to detect interspecific interactions. The neutral model of Caswell (1976) will be taken as an example.
4. Use of species-area relationship to distinguish between founder-controlled and dominance-controlled communities (Lepš and Štursa 1989).
5. Combining spatial pattern with supplementary information about particular individuals (their vitality, size, etc.).

It is concluded that evidence obtained from manipulative experiments is usually stronger than that deduced from observed patterns. Nevertheless, the analysis of patterns is necessary for suggesting proper hypotheses to be tested. The more appropriate auxiliary variables are included in the analysis of pattern, the more are the results suggestive of underlying mechanisms.

Introduction

Attempts to estimate the underlying mechanisms from observed patterns are numerous. Opinions about the reliability of these estimates differ considerably. In classic works and textbooks of quantitative ecology (Greig-Smith 1952, Kershaw 1974), the spatial pattern of a population (measured as the intensity of clumping) or the frequency of statistically significant "interspecific associations" have been considered to be measures of biological organization. For example, Kershaw (1974, p. 57, repeated in Kershaw and Looney 1985) writes: "Interactions between species or individuals of the same species can be conveniently measured by means of a χ^2-test or by an analysis of the 'pattern' of species distribution

1

respectively." In contrast, Shipley and Keddy (1987) clearly state that: "Hypotheses of pattern should be tested using inferential statistics. Hypotheses of mechanisms should be tested by experimental manipulations, making sure not to equate correlation with causation." This statement is well supported in their paper and it is clear that the evidence obtained from manipulative experiments is much more reliable than that based on observed patterns. Unfortunately, manipulative experiments are not always feasible or practicable. They may be extremely expensive or time-consuming. The aim of this paper is to critically examine the possibilities for estimating underlying mechanisms by studying patterns, with special attention paid to spatial patterns. The aim of this paper is neither to review the particular methods of pattern description nor to evaluate their statistical validity. The paper focuses on the interpretation of results. Thus methods of description will be accepted (usually from authoritative textbooks) without any discussion. The following cases will be discussed in detail:

1. The spatial pattern of a single population and interactions among individuals.
2. The spatial arrangement of two populations (statistical "interspecific associations") and interactions among populations.
3. Analysing the distribution of population abundances and comparison with a neutral model to detect interspecific interactions. The neutral model of Caswell (1976) will be taken as an example.
4. Use of species-area relationship to distinguish between founder-controlled and dominance-controlled communities (Lepš and Štursa 1989).
5. Combining spatial pattern with supplementary information about particular individuals (their vitality, size, etc.).

The mechanisms may be deduced either by comparing a single observed state with a state generated (predicted) by a neutral model (i.e. a model in which interactions we intend to demonstrate are not included), or from differences between two or more states in a time sequence.

Spatial pattern of a single population

There are several ways to describe spatial pattern of a species (see e.g. Kershaw 1974, Pielou 1977, Ripley 1987). As a neutral model, the "random pattern" is considered, where individuals are distributed randomly and independently. There are many ways to compare this neutral model with the observed distribution of individuals. Generally, the approaches fall into two basic groups: so called distance methods, based on measurements of distances between individuals, and methods based on counts in quadrats. Theoretically developed, but seldom used is the analysis of two-phase mosaic (Pielou 1977).

Pattern is described by its scale and intensity (size of the departure from randomness). In the aggregated pattern, the probability of finding an individual is increased by the presence of another individual. As a result, an aggregated pattern has been considered as an indicator of positive interactions among individuals and regular (uniform) pattern as an indicator of negative interactions. However, there are many causes of clumping (most of which are reviewed in Kershaw 1974); the most important are environmental heterogeneity, type of dispersal, and competition with other species (sometimes called environmental, morphological, sociological pattern). On the other hand, the tendency toward regularity

2

may be almost entirely ascribed to competition among neighbouring individuals. The influences of particular factors act on various scales and often cancel one another. The pattern of each population is influenced by historical events (for example, the factors influencing the establishment of an individual might have no influence on an adult; the distribution of anemochorous seed is influenced by microrelief formation). Consequently, the resulting intensity of spatial pattern is influenced by many different factors and it may hardly be considered as a measure of a single one. The causes of aggregation are statistically indistinguishable (Ripley 1987). For example, as shown by the simulation study by Lepš and Kindlmann (1987), a nearly random spatial pattern may result from selfthinning of an initially aggregated population due to strong competition between neighbouring individuals.

Different factors express their effect on different scales. For example, it is often found that individuals are evenly distributed on a small scale (due to competition among neighbours) and aggregated on larger scales (due to environmental heterogeneity). Consequently, pattern analysis that considers various scales is much more informative and suggests much more about underlying mechanisms, particularly when it is accompanied by analysis of environmental factors (Kershaw 1961). It should be noted that the classic distance methods (Hopkins and Skellam 1954, Clark and Evans 1954) mix together the influence of pattern on various scales, with small scale phenomena given highest weight. This is caused by measuring only distance to the nearest neighbour. Recently, methods have been developed (Ripley 1987) using distances to all other individuals in the population; in these, various scales may be considered. However, they are practicable mainly for analysis of digitized maps.

Whereas the interpretation of "snapshot" analysis of pattern is difficult, much more may be said from repeated observations. In particular, changes toward regularity may be considered as a good indicator of negative interactions between neighbouring individuals (usually competition, see Lepš and Kindlmann 1987). Increase in clumping suggests positive interaction (Vacek and Lepš 1987). However, even in this case, it is a suggestion only — some other factors may play a role, such as changes in sensitivity to environmental factors with individual age, or uncontrolled changes in environmental factors.

Interspecific associations

As in the case of single species pattern, there are two groups of methods for determination of "interspecific associations". The first group is based on presence or counts of individuals in a set of plots (usually quadrats). The second is based on comparison of the relative position of individuals of species or on their distances (testing for segregation according to Pielou 1977).

We shall consider the determination of interspecific associations on the basis of presence-absence data (described in textbooks of quantitative ecology, e.g. Kershaw 1974). However, very similar reasoning may be applied to assessment based on quantitative data. In this case, the presence of the two species is noted in a set of (usually randomly placed) quadrats and the number of common occurrences is compared with that expected under the hypothesis of independence of the two species. The significance of departure is tested using a 2×2 contingency table by a χ^2-test. A significant result of the test is labelled as a positive or negative interspecific association. Usually, all the species are recorded and the procedure is repeated for all possible pairs of species or for species with frequencies higher than a certain threshold. The percentage of significant results has been considered to be a measure of community organization (Greig-Smith 1952). There are several problems with this procedure: first, there are problems of a statistical nature — the repeated test for all pairs of species leads to multiple inference from a single data set and increased probability of type I error. Analysis of multiway contingency tables requires a very high number of sampling units when the number of species exceeds four. Limited information on prevalence of positive or negative associations may be obtained from species counts in sampling units (Barton and David 1959). Within a single pair, the strength of test varies with frequencies of particular species and is particularly low when the frequency is too high or too low. Consequently, the percentage of species pairs exhibiting a significant association is considerably affected by the heterogeneity of the plot and by the distribution of species frequencies. Pielou (1969) has shown the possibility of detecting "random association" as a sampling artefact. However, more important difficulties appear with interpretation. Whereas the direct trophic relationships (like those of parasitic plants and their hosts) are usually studied by other methods, the association is usually considered to be active when one species modifies the environment in a way supporting (positive association) or suppressing (negative association) the other species. The passive association is caused by concordant (positive association) or discordant (negative association) response to some third factor. The third factor is usually interpreted as an environmental factor, but it may well be the rest of the community; two species may be found to exhibit a positive association simply because their mutual competitive relation is weaker than that with other plants. For example, in the steppe grasslands, the spring therophytes are usually found in common in gaps between tussocks of perennial grasses. Conflicting interpretations of interspecific association are possible (according to what we consider to be a cause and what we consider to be a consequence): (1) The (active) negative association is a consequence of competition between species - consequently the species pairs exhibiting the negative association are those with the strongest competitive relations. (2) The (passive) positive association is a consequence of concordant response to an environmental factor; the species exhibiting the positive association have the most similar niche and so they are the strongest competitors.

There is no way to distinguish the causes of associations from the 2×2 table. The omission of the d-field (e.g. Fager 1957) is unjustified (see Pielou 1977). Similarly, the 2×2 table does not provide information on the mutual influence of the two species (i.e. which species is influencing and which is the influenced one). All the two-sided indices (e.g. Ipatov et al. 1974) show that the more influential species is the more frequent one — Fig. 1 provides a simple counterexample.

It has been shown that interspecific associations change with the size of the plot of the sampling unit (Kershaw 1974). For example, two species may exhibit negative association on a small scale due to competitive exclusion and positive association on a larger scale due to concordant response to some environmental factor. So the common pattern analysis of the two species (Kershaw 1961) shows the mutual spatial relationship better and is more suggestive about their causes than any observation on a single scale. Even more suggestive

4

Fig. 1. A simple example, where a herbaceous plant depends on two species of tree. The herbaceous plant has greater frequency than either of the two trees and consequently, any two-sided index shows that the trees depend on the herb.

results are obtained when underlying environmental factors are included in the analysis.

In conclusion, analysis of interspecific associations describes the spatial relationships of populations and, in many cases, is suggestive about their causes. The percentage of interspecific associations (or some similar index) may be regarded as a measure of spatial heterogeneity or spatial diversity (a measure of departure from random independent dispersion), but by no means as a measure of functional organisation. Hence it is unjustified to deduce the validity of the individualistic community concept from the minimum interspecific associations found in climax communities (as in Greig-Smith 1952, or Kershaw and Looney 1985). Similarly, attempts to estimate the community matrix (or other dynamic parameters of multipopulation models) on the basis of "interspecific associations" are unsound. Similar conclusions were drawn by Hastings (1987).

The methods based on distances between individuals are described by Pielou (1966) for many species (the pattern diversity) and by Pielou (1974, p. 227) or Ripley (1987) for two species. The approach is based on the composition of nearest neighbour pairs or groups or on the distribution of monospecific and mixed pairs of individuals. As shown in the above mentioned papers, the segregation is influenced by both the initial pattern of studied species and the interactions among individuals (both inter- and intraspecific). However, some conclusions about mechanisms may be suggested when a time-sequence of observations is available. In the above mentioned case, a decrease in segregation (increase in the pattern diversity) in the course of self-thinning suggests the influence of intraspecific competition.

5

Distribution of species abundances and analysis of neutral models

Attempts to determine the degree of biological organization (the strength of biological interactions) from a single "relevé" (i.e. abundances of all species present on a site) are among the most ambitious projects in community ecology. The usual procedure compares the observed state with that predicted by a neutral model; that of Caswell (1976) is probably the most popular (see references in Lambshead and Platt 1988). In this case, the observed diversity is compared with that predicted by the neutral model. However, the analysis given in this paper will show that a single relevé does not contain sufficient information to do this. Here I will first develop general arguments about neutral models and then will focus on the model of Caswell (1976). The model III adapted from the neutral allele model of Ewens (1972) which is used in most calculations will be discussed.

The procedure should be performed in two steps: first, the neutral model may be rejected and second, if so, the departure from the neutral model is interpreted as a measure of biological organization. It seems meaningless to test the neutral model in communities where there are direct trophic links, because the "no interaction" assumption may be rejected *a priori*. The interpretation of the size of departure from predicted diversity as a measure of biological organization is problematic, too. For example, the selective and non-selective grazing may have contradictory effects on diversity and as a final result, they may negate each other.

In each neutral model, two assumptions are included — (1) there are no interactions among populations and (2) there are no differences among population responses to the environment (i.e. there are no differences among carrying capacities of the environment for particular species). Consequently, the differences in population abundances are purely a matter of chance. The second assumption is not always explicitly stated, but is always included. (Note that in the original genetic model of Ewans, both the corresponding assumptions are feasible: the allele is selectively neutral, if it changes neither an individual's competitive ability nor its response to environment.) The rejection of the neutral model means that either one of the assumptions is not valid. There is no *a priori* reason to conclude that the first assumption is violated. The relevé does not contain any information which would enable one to reject the assumption that species abundances are determined solely by the environment (i.e. that abundance of each species corresponds to its carrying capacity); in other words, that the species would attain the same abundance if grown in the site alone. A habitat usually provides a (physiological) optimum for some species, whereas it is only suboptimal or marginal for others.

Competition proceeds on the "individual with individual" level; similarly, the model of Caswell (1976) is based on individuals (i.e. on the random exchange of individuals regardless of species). Hence, the model does not describe a community without interactions among individuals, but one where the intraspecific and interspecific interactions (among individuals) are the same for all species. (The hypothesis that there are no interactions among individuals of a population may be rejected *a priori*). This interpretation follows immediately from the analogy with the original genetic model - individuals differing in a locus do compete, but in the same way as individuals with the same allele. Similar points have been made by Ugland and Gray (1983), who state that fixed community size means competition. The response of Caswell (1983) did not reject their points. He states that

6

"Since Models II and III assume a fixed number of individuals, and Model I does not, nothing in the conclusions can possibly be an artifact of this fixed community size." The reliance on equivalence of the three models is unsound. If several models, some of which include interspecific interactions, give similar predictions of distribution of species abundances or of diversity, then the prediction may not be used as a yard-stick for detection of interactions.

In conclusion, the distribution of species abundances and their comparison with neutral model predictions is able neither to detect nor to measure the biological interaction within a community. The interpretation of departures is ambiguous.

On the other hand, it has been found that certain types of distribution (e.g. broken-stick, logarithmic, lognormal, see Whittaker 1975) are typical of certain types of communities or of certain successional stages, and may be suggestive in some cases, particularly when used on a comparative basis. But the mechanisms leading to a particular distribution may differ considerably; for example, Cohen (1968) has shown that the distribution predicted by the broken stick model may result also from other underlying models, and consequently this distribution may not be regarded as evidence for any of them.

Diversity also has been used to measure community stability. The first attempt was by MacArthur (1955). However, no unique relationship between diversity (or distribution of species abundances) and stability was found (Pimm 1986), and so it is unjustified to measure stability by community diversity. Particularly misleading is an approach when no operational definition of stability is provided and diversity (with other structural parameters) is used to quantify this fuzzy concept, as was attempted e.g. by Ružička et al. (1983).

Species–area relationship and the type of competitive interactions

Yodzis (1978) distinguished two types of communities of sedentary organisms with regard to the type of competitive network — founder-controlled and dominance-controlled. Roughly speaking, in founder-controlled communities, the species occupying a gap first (the founder) remains there, whereas in a dominance-controlled community it may be outcompeted by another species. It is clear that founder and dominance control are the two extreme points of a continuum. Based on a set of simulation experiments with a reaction-dispersal model, Yodzis predicted differences between the species-area relationships of the two types of communities. When considering the relationship $S = c.A^z$, where S is the number of species, A is area and c and z are parameters, the value of the z-parameter should be higher in founder-controlled communities. The species-area relationship on the within-community scale (not influenced by habitat heterogeneity) is considered.

Lepš and Štursa (1989) examined species-area relationships of two sets of communities and found that the relationship, particularly the parameter z, changes in a predictable way in the course of succession and on an environmental gradient. The changes may be partially accounted for by the distinction between founder-controlled and dominance-controlled communities (Yodzis 1978). According to the model of Yodzis (1978), this distinction is caused by differences in niche differentiation between species. Could the form of the species-area relationship thus be used to estimate the underlying competitive relationships and even the degree of niche differentiation?

7

The species-area relationship is influenced also by other factors, namely by the available species pool. The species pool may be limited by dispersal constraints, or by constraints superimposed by the physical environment (not all the species propagules which are present are able to grow in a given environment). These constraints usually cause a decrease in the z-value in the species-area relationship. It seems that the high value of z (in our case about 0.3 and higher) may be taken as very suggestive of founder-control, whereas low values may have several causes. To get a better idea, the information should be combined with information on life histories of species present. In contrast, the relationship between the degree of niche differentiation and founder/dominance-control has not been tested experimentally and the use of species-area relationship for estimating extent of niche differentiation would be unsound.

Combining spatial pattern with supplementary information about particular individuals

Pielou (1974, p.252) described a method for detection and crude measurement of competition using spatial pattern. The method involved correlating the distance from a tree to its nearest neighbour with the sum of their circumferences (both variables were transformed to logarithms). A positive correlation was considered to be a sign of competition. In a multispecies stand, the correlations of single species and mixed species pairs may be compared. Similar reasoning was used by Vacek and Lepš (1987) to compare changes in the degree of defoliation caused by air pollution stress with distance to the nearest neighbour. In some cases, the evidence for "ecological sheltering" was found. Malik *et al.* (1976) compared the spatial pattern of individuals of *Atriplex vesicaria* of various sizes (ages) and found that the young ones are more clumped, which is the indication for competition among neighbours. Ishizuka (1984) used the comparison of the spatial pattern of centres of crowns with the pattern of stems in various layers for the study of processes influencing crown distribution.

These methods seem to be the most promising ones, particularly when a time series of observations is used. For example, the total size of two neighbours is influenced not only by their competition. The size of the yearly increment adjusted to the total tree size may be a better indicator of competition. (Tree ring analysis may provide useful information.) Generally, analyses that include a greater degree of appropriate information about individuals will give more constructive results.

Pattern recognition and experimentation

As noted by Tillman (1988), the central goal of ecology is to understand the causes of the patterns we observe in the natural world. However, to do this, the patterns have to be sufficiently described. It is obvious that properly designed experiments provide much stronger inference about underlying mechanisms than observed patterns themselves. Experiments are usually done to test some hypothesis about underlying mechanisms. For a test to be strong enough, the hypothesis must be stated as precisely as possible. Moreover, experimental tests, particularly those carried out in the field, are laborious and time-consuming and have many limitations (see Diamond 1986). As noted by Haila (1988), science is not advanced through planning rigorous tests of patently unrealistic hypotheses. The analysis of pattern is essential for suggesting hypotheses that are both testable and interesting. It is necessary to search for repeatable patterns, for similarities from habitat to habitat. It is impracticable to carry out experiments in all these habitats. The strength of evidence based on these observed patterns differs among particular cases; reliable evidence is obtained when it is known beforehand that some pattern is correlated with some independently (experimentally) measured dynamic characteristics. Weak evidence is usually obtained from comparison of an observed pattern with a pattern generated by a neutral model; all the neutral models we examined are patently unrealistic and/or are very sensitive to factors other than the mechanisms that the models are designed to detect. The use of the size of departure from the pattern predicted by a neutral model is also problematic — particularly when the effects of various mechanisms on the resulting pattern negate one another. The danger of misinterpretation increases when some characteristic of the observed pattern is used as a direct estimate of some dynamic parameter of underlying mechanisms, particularly when the estimated parameter is undefined or poorly defined (as with a measure of biotic organization, stability, etc.). Sometimes, the conclusiveness differs even with the direction of the departure. For example, whereas a clumped spatial pattern is very common in nature and itself says nothing about possible causes, a uniform pattern is relatively rare and is highly suggestive of competition among neighbouring individuals.

It should be stressed that, whereas the departure may be suggestive of a particular mechanism, the agreement with neutral model prediction provides extremely weak (if any) evidence for absence of a mechanism. For example, neither random nor clumped patterns indicate that neighbouring individuals do not compete.

More meaningful evidence may be obtained from a time sequence of observations. In some cases, comparison between sites may also be useful. When the spatial and temporal comparisons are combined, the evidence may be very suggestive (e.g. when we compare the development of spatial pattern of two populations differing in initial density). Such comparisons may be considered to be "natural experiments"; however, in comparison with manipulative experiments, they lack the advantage of experimental design. In the above example, the differences in initial density may be caused by an uncontrolled factor.

Very compelling evidence is obtained when the information about spatial pattern is combined with additional information, particularly when such a combined observation is carried out in a time sequence.

It is obvious that pattern analysis serves mainly to suggest hypotheses about underlying mechanisms, which can be tested through manipulative experiments. Nevertheless, the distinction between uses of pattern analysis and manipulative experiments is not absolute. Depending on the spatial and temporal scales of the problem, manipulative experiments are considerably limited. Reliance on natural experiments is often unavoidable (Diamond 1986). The relative representation of the two approaches depends on the "maturity" of research in a certain area (Loehle 1987). Experimental studies are implemented after plausible hypotheses have been suggested by observed repeatable patterns.

It may be concluded that mechanisms can be suggested on the basis of observed patterns, but they cannot be tested. In no case does the size of the departure of observed pattern from randomness (from the prediction of neutral model) allow for measuring the intensity of biological interactions. The limited interpretability of observed patterns does not deny the general usefulness of pattern analysis.

Acknowledgements. Thanks are due to Nancy Moran for her kind linguistic help and for comments on the manuscript, and to Milan Straškraba for critical reading of the first draft of the paper.

References

Barton D.E. and F.N. David (1959): The dispersion of a number of species.—J.R.Statist.Soc.B. 21: 190–194.

Caswell H. (1976): Community structure: a neutral model analysis. —Ecol. Monogr. 46: 327–354.

Caswell H. (1983): Reply to a comment by Ugland and Gray.—Ecology 61: 194–204.

Clark P.J. and Evans F.C. (1954): Distance to the nearest neighbor as a measure of spatial relationships in populations.—Ecology 35: 445–453.

Cohen J.E. (1968): Alternate derivations of a species-abundance relation.—Am. Nat. 102: 165–172.

Diamond J. (1986): Overview: Laboratory experiments, field experiments, and natural experiments.—In: Diamond J. and Case T.J.[ed.], Community ecology. Harper and Row, New York. pp. 3–22.

Ewens W.J. (1972): The sampling theory of selectively neutral alleles.—Theoret. Pop. Biol. 3: 87–112.

Fager E.W. (1957): Determination and analysis of recurrent groups.—Ecology 38: 586–595.

Greig-Smith P.(1952): Ecological observations on degraded and secondary forest in Trinidad, British West Indies.—J. Ecol. 40: 316–330.

Haila Y. (1988): The multiple faces of ecological theory and data.—Oikos 53: 408–411.

Hastings A. (1987): Can competition be detected using species co-occurrence data?—Ecology 68: 117–123.

Hopkins B. and Skellam J.G. (1954): A new method for determining the type of distribution of plant individuals.—Ann. Bot. 18: 213–227.

Ipatov V. S., Samojlov J.I. et Tarchova T.N. (1974): Dvustoronnyj koeficient mezvidovoj soprjazenosti. [The duplex coefficient of specific correlation.—Bot. Zhurn. 59: 1596–1602.

Ishizuka M. (1984): Spatial pattern of trees and their crowns in natural mixed forests.—Jap. J. Ecol. 34: 421–430.

Kershaw K.A. (1961): Association and co-variance analysis of plant communities.—J.Ecol. 49: 643–654.

Kershaw K.A. (1974): Quantitative and dynamic plant ecolog—2nd ed. Edward Arnold, London.

Kershaw K.A. and Looney J.H.H. (1985): Quantitative and dynamic plant ecology.—3rd ed. Edward Arnold, London.

Lepš J. and Kindlmann P. (1987): Models of the development of spatial pattern of an even-aged plant population over time. —Ecol. Modelling 39: 45–57.

Lepš J. and Štursa J. (1989): Species-area curve, life history strategies, and succession: a field test of relationships. —Vegetatio 83: 249–257.

Loehle C. (1987): Hypothesis testing in ecology: psychological aspects and importance of theory maturation.—Quart. Rev. Biol. 62: 397–409.

Malik A.R., Anderson D.J. and Myerscough P.J. (1976): Studies on structure in plant communities. VII. Field and experimental analyses of *Atriplex vesicaria* populations from the riverine plain of New South Wales.—Austr. J. Bot. 24: 265–280.

Pielou E.C. (1969): Association tests versus homogeneity tests: their use in subdividing quadrats into subgroups.—Vegetatio 18: 4–18.

Pielou E.C. (1974): Population and community ecology.—Gordon and Breach, New York.

Pielou E.C. (1977): Mathematical ecology.—Wiley, New York.

Ripley B.D. (1987): Spatial point pattern analysis in ecology.—In: Legendre P. and Legendre L. [ed.], Developments in numerical ecology. Springer, Berlin. pp. 407–429.

Ružička M., Jurko A., Kozová M., Žigrai F. et Svetlosanov V. (1983): Evaluation methods of landscape stability on agricultural territories in Slovakia.—Ekologia (ČSSR) 2: 225–253.

Shipley B. and Keddy P.A.(1987): The individualistic and community concepts as falsifiable hypotheses.—Vegetatio 69: 45–55.

Tilman D. (1988): Plant strategies and the dynamics and structure of plant communities.—Princeton University Press, Princeton, N.J.

Ugland K.I. and Gray J.S. (1983): Reanalysis of Caswell's neutral models.—Ecology 64: 603–605.

Vacek S. and Lepš J. (1987): Changes in the horizontal structure in a spruce forest over a 9-year period of pollutant exposure in the Krkonoše Mountains, Czechoslovakia.—Forest Ecol. Manage. 22: 291–295.

Whittaker R.H. (1975): Communities and ecosystems.—2nd ed. Macmillan, New York.

Yodzis P. (1978): Competition for space and the structure of ecological communities.—Springer, Berlin.

Krahulec F., Agnew A.D.Q., Agnew S. & Willems J.H. [eds.]: *Spatial processes in plant communities.* pp. 13–22.

Interspecific associations in old-field succession

Jan Lepš[1] and Václav Buriánek[2]

1) *Department of Biomathematics, Biological Research Centre, Branišovská 31, CS-370 05 České Budějovice, Czechoslovakia.*
2) *Research Institute for Forestry and Game Management, Jíloviště - Strnady, CS 255 01 Zbraslav, Czechoslovakia*

Keywords: Community structure, Competition, Interspecific associations, Life histories, Old-field, Phenology, Succession.

Abstract. Interspecific associations were determined on the basis of quantitative (cover) and qualitative data in three old-fields of different age and in nearby natural forest in the Bohemian Karst, Czechoslovakia. No consistent trend was found in changes of the frequency of significant associations with the successional age of the field. The pattern of species associations was analysed with respect to phenological similarity of species and with respect to life histories of species. The hypothesis that phenologically similar species compete more than phenologically dissimilar species and consequently are spatially segregated is only weakly supported by the data from the oldest field. Except in the youngest field, differences were found between associations of species belonging to various groups. The associations among annuals were mostly positive; the strongest negative associations were found among perennial grasses. The frequency of significant associations may be regarded neither as a measure of biological organisation, nor as a measure of the competition strength.

Introduction

Statistically determined associations between species are considered to be an important characteristic of a plant community (Kershaw and Looney 1985). Changes in the pattern of interspecific associations in the course of succession or in variously disturbed communities have been investigated (e.g. Greig-Smith 1952, Kellman 1969, O'Connor and Aarssen 1987). In early interpretations, the frequency of statistically significant associations was considered to reflect the degree of biological organization of a community, and its decrease in 'stable' or late successional communities was considered to support the individualistic concept of plant communities (Kershaw 1973). In a recent paper (O'Connor and Aarssen 1987), the pattern of associations is related to the competitive structure.

The aim of this paper is to answer following questions:

1. Are there any changes in the pattern of interspecific associations in the course of secondary succession?

2. Are there any relationships between ecological and/or morphological similarity of species and their spatial relationships?

Within the second question, two hypotheses were tested.

(a) Species with similar phenology are supposed to compete more than species with differing phenology and as a consequence they should be spatially segregated. A negative correlation is expected between a measure of phenological similarity and a measure of interspecific association. This assumption was proposed by Hurlbert (1969).

(b) Strong competitors are supposed to be spatially segregated (to be negatively associated); weak competitors are supposed to be found together in the "gaps" between the strong competitors and are expected to be positively associated.

Material

The structure of vegetation in three old-fields differing in age and in natural forest nearby was investigated in the framework of a larger project on old-field succession in the Bohemian Karst (Osbornová *et al.* 1990). The Bohemian Karst is a limestone area in Central Bohemia, Czechoslovakia. The climate is summer-warm and dry (annual averages 8°C and 370 mm of precipitation). The natural vegetation in the area consists of thermophilous forests dominated by *Quercus petraea*, *Q. pubescens* and *Carpinus betulus*, and species-rich rocky steppes, particularly on south-facing limestone slopes.

Two successional seres were distinguished in old-field succession — the xeric one and the mesic one. The investigated fields belong to the xeric sere (intensively studied fields labelled X1, X2 and X3 in Osbornová *et al.* 1990). All three fields are situated in the same position on the tertiary Berounka river terrace (mainly acidic sediments on limestone parent rock) near the village of Srbsko. The youngest field (size 15 × 50 m) was sampled in the first four years after abandonment (1979 to 1982). The last crop was wheat, the field was ploughed after harvesting. The medium aged field (15 × 60 m) was sampled at the age 7 and 11 yrs (1976,1980) and the oldest field (40 × 30 m) at the age of ca. 55 yr (1975). For comparison, some investigations were carried out in the ground layer of a forest of a natural species composition on limestone approx. 4 km away from the investigated old-fields.

In the first year, the old-field vegetation consisted mainly of annual weeds. The most abundant was *Papaver rhoeas*, other species were *Fagopyrum convolvulus*, *Medicago lupulina*, *Galium aparine* and *Veronica* spp. In subsequent years, annuals were subsequently replaced by biennials and perennials (mainly *Artemisia vulgaris*, *Agropyron repens*, and *Daucus carota*, *Taraxacum officinale*, *Achillea millefolium*). *Artemisia* and *Agropyron* were abundant in the 4-th year of the youngest field and were dominant in the medium aged field about its 7-th year. Then they lost ground to other species, mainly perennial grasses. The oldest stadium was without any directional change in the last years (1975 to 1988). It was dominated by narrow-leaved perennial grasses *Festuca rupicola* and *Poa angustifolia*. The forest ground layer was without any apparent dominant. It was species rich with mainly broad-leaved perennials, for example *Lathyrus vernus*, *Sanicula europaea* and *Asarum europaeum*.

Within the larger project, manipulative perturbation experiments were carried out in the medium-aged and the old plots. In particular, increase of biomass of unaffected

14

species after application of selective herbicides (Michálek, Buriánek and Hadincová 1990) has demonstrated influence of competition. This enabled comparison of evidence obtained from manipulative experiments and from analysis of distributional patterns.

Methods

The interspecific associations at the 1 m^2 scale were determined in the first four years of the youngest field, at age 7 yrs of the medium aged field, in the oldest field and in the forest ground layer once (in 1975). In each plot at each sampling date, forty 1 m^2 quadrats were analysed by the point quadrat method (Goodall 1952), 100 pins in a square grid in each quadrat. In this way, estimates of cover of all species were obtained. For the youngest field, the data are part of those used by Lepš (1987). The interspecific associations were determined on the basis of correlation coefficient r for all pairs of species with frequency higher than 10%. The frequency of statistically significant associations was determined (as a ratio of number of significant associations and number of all possible pairs of species with frequency higher than 10%) for all associations at $P < 0.05$, $P < 0.01$ and for positive associations at $P < 0.05$. Based on the same data, the heterogeneity of each plot (β-diversity) was determined as the difference between the diversity of pooled sample of all quadrats from the plot (γ-diversity) and average diversity in the quadrat (α-diversity). The diversity index H was computed according to the Shannon formula (logarithm base 2).

The interspecific associations on a small scale were determined once in each field (younger one at age 2 yrs, medium at 11 yrs, the oldest one at approx. 60 yrs). In each field, 150 squares 0.1 × 0.1 m were laid down and the root presence was noted in each. The interspecific associations were determined by usual analysis of 2 × 2 tables for all pairs of species. Species which appeared in 5 and fewer quadrats were excluded from the analysis. The significance of the association was tested by a χ^2-test and the strength of the association was expressed by the V coefficient:

$$V = \frac{(ad - bc)}{\sqrt{mnrs}} \tag{1}$$

where a, b, c and d are values of fields of 2 × 2 table and m, n, r and s are corresponding marginal totals.

On the medium aged and on the old-fields, the phenology of particular species was followed. At approximately two week intervals, the phenological phase of each species was determined. For computing the phenological similarity, only the data on flowering were used. The flowering in time t, $F(t)$ was assigned to equal 2 when most of the population was in flower, 1 if part of the population was in flower and 0 if there was no flowering. The phenological similarity of species i and j, $PS_{i,j}$ was then defined

$$PS_{i,j} = \frac{2 \sum_{t=1}^{n} \min(F_i(t), F_j(t))}{\sum_{t=1}^{n} F_i(t) + \sum_{t=1}^{n} F_j(t)} \tag{2}$$

n is total number of time intervals sampled. The relationship between the association of two species V and their phenological similarity PS was evaluated by the correlation coefficient r. If there is a spatial exclusion of phenologically similar species, r is expected to

15

be significantly lower than zero. However, corresponding degrees of freedom (to estimate the variance of r and to test the null hypothesis: $r \geq 0$) are difficult to determine - particular values of coefficients are *not* independent variables. Consequently, the significance of the departure was tested using Monte Carlo permutation test. One of the matrices was subjected to random permutation of species and the correlation coefficient was computed. The significance was estimated from the proportion of values of r smaller than the original value of r. In most cases, 500 random permutations were used. The relationship is expected to be more pronounced in species competing with each other. Consequently, the same procedure was carried out with species pairs exhibiting the negative association only.

Species were divided into four (groups of) life forms: annuals, biennials, perennial grasses, perennial forbs. (No perennial grasses with sufficient frequency in the youngest field and no biennials in the oldest field were found). The average value of the coefficient V was computed for each combination of species life forms (i.e. annual - annual, annual - biennial, annual - perennial grass, etc.). The number of species pairs within a group is $\binom{n}{2}$, i.e. $\frac{(n-1)n}{2}$, between groups $n_1 . n_2$ (n is number of species in respective group). For the same reasons as above, the Monte-Carlo permutation test was used to test the hypothesis that the value of the coefficient does not depend on the combination of life forms. (The appropriate number of life forms was randomly assigned to species). Two different test-criteria were used: the F-value for one-way ANOVA and absolute value of difference between the highest and the lowest group average. 500 random permutations were used.

Results

On the 1 m^2 scale, there is no consistent trend in the number of significant interspecific associations in the course of succession (Fig. 1). The percentage of significant associations is low (close to the number of associations resu ing from Type I error). Similarly, no consistent trend was found in the β-diversity (horizontal heterogeneity) of the plot. The forest ground layer has the lowest percentage of significant associations and very low β-diversity. Moreover, the structure of associations is difficult to interpret. As an example, the structure of interspecific associations in the medium aged and in the oldest fields are presented in Figs 2. and 3.

Similarly, neither consistent trend, nor striking differences in the number of significant associations determined on the basis of presence-absence were found (Table 1). In no case significant correlation was found between the strength of association and phenological similarity (Table 2). In the medium-aged field, the correlation was positive but very close to zero. In the old-field, the correlation was negative, as expected, but not significant. When only pairs exhibiting negative association were considered, the correlation was suggestive ($P = 0.065$). It suggests that in the older field the horizontal structure of the community might be shaped by competition more than in the younger one, or that in the older field there are some species pairs that are able to avoid competition by phenological differentiation. However, the evidence is weak.

The differences between average values of V in groups of species pair types are significant in the medium-aged and old fields and non-significant in the young field (Tables 3, 4, 5). In both cases where significant differences were detected, the average association

16

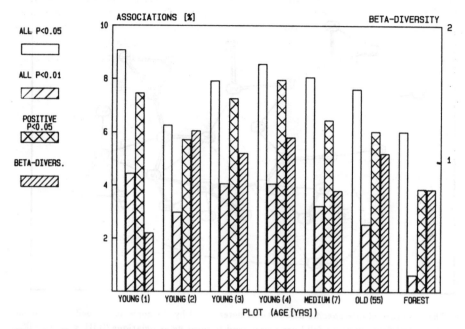

Fig. 1. Percentage of significant associations determined by correlation coefficients and β-diversity in old-fields of various ages and in the forest.

Table 1. Numbers of species included in the analysis and the percentage of significant associations detected by the χ^2 analysis of 2x2 tables.

field	number of species	percentage of significant associations	
		for $P < 0.05$	for $P < 0.01$
young	26	3.08	1.23
medium aged	24	6.16	2.17
old	23	5.50	1.58

between pairs of annuals is positive and the average association between pairs of perennials is negative. The annuals are found together in gaps between the warp of perennial species. The averages are low in absolute values - however, note that for 150 sampling units $|V| > 0.16$ means significant association. The maximum possible value of V is determined by marginal totals (i.e. by frequencies of the species) and is low for a non-symmetrical table. Again, it seems that the differentiation of species competitive abilities increases with increasing field age. In the young field (age at sampling 2 yrs), the warp of perennial species was not present and the majority of species were annuals.

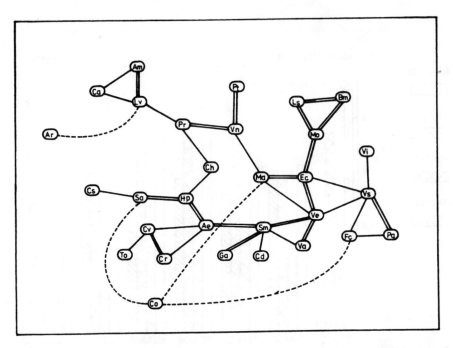

Fig. 2. The structure of interspecific associations determined by the correlation coefficients in the medium aged field. Straight full lines correspond to positive associations ($0.01 < P < 0.05$), double straight lines to positive associations ($P < 0.01$) and broken curves to negative associations ($P < 0.05$). Meaning of abbreviations: Ae – *Aegopodium podagraria*, Am – *Achillea millefolium*, Ar – *Agropyron repens*, Bm – *Bromus mollis*, Ca – *Cirsium arvense*, Cd – *Carduus acanthoides*, Ch – *Chenopodium album*, Co – *Convolvulus arvensis*, Cr – *Consolida regalis*, Cs – *Centaurea scabiosa*, Cv – *Coronilla varia*, Ec – *Eryngium campestre*, Fc – *Fallopia convolvulus*, Ga – *Galium aparine*, Hp – *Hypericum perforatum*, Ls – *Lactuca serriola*, Lv – *Linaria vulgaris*, Ma – *Myosotis arvensis*, Mo – *Melilotus officinalis*, Pa – *Poa angustifolia*, Pr – *Papaver rhoeas*, Pt – *Potentilla argentea*, Sa – *Silene alba*, Sm – *Stellaria media*, To – *Taraxacum officinale*, Va – *Viola arvensis*, Ve – *Veronica arvensis*, Vi – *Vicia angustifolia*, Vn – *Verbascum nigrum*, Vs – *Veronica sublobata*.

Table 2. Values of correlation coefficients r between the V-values and phenological similarity for all V values and for negative V values only and corresponding significances P (test of the null hypothesis $r \geq 0$).

	All V-values		Negative V-values	
	r	P	r	P
medium aged field	0.038	0.740	0.102	0.880
the oldest field	-0.047	0.255	-0.122	0.068

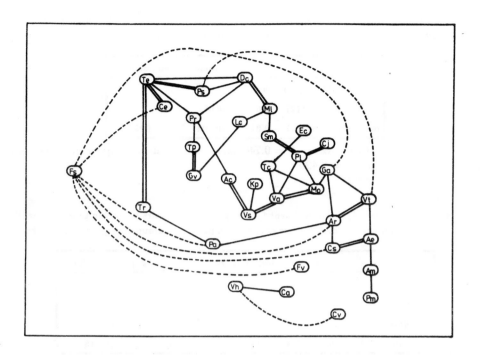

Fig. 3 The structure of interspecific associations determined by the correlation coefficients in the oldest field. Straight full lines correspond to positive associations ($0.01 < P < 0.05$), double straight lines to positive associations ($P < 0.01$) and broken curves to negative associations ($P < 0.05$). Meaning of abbreviations: Ac – *Asperula cynanchica*, Am – *Achillea millefolium*, Ar – *Agropyron repens*, Ae – *Arrhenatherum elatius*, Ca – *Convolvulus arvensis*, Ce – *Cerastium arvense*, Cj – *Centaurea jacea*, Cs – *Centaurea scabiosa*, Cv – *Coronilla varia*, Dc – *Dianthus carthusianorum*, Ec – *Euphorbia cyparissias*, Fs – *Festuca rupicola*, Fv – *Fragaria viridis*, Ga – *Galium album*, Gv – *Galium verum*, Kp – *Koeleria pyramidata*, Lc – *Lotus corniculatus*, Ml – *Medicago lupulina*, Mo – *Melilotus officinalis*, Pa – *Poa angustifolia*, Pl – *Plantago lanceolata*, Pm – *Plantago media*, Pr – *Potentilla reptans*, Ps – *Pimpinella saxifraga*, Sm – *Sanguisorba minor*, Tc – *Trifolium campestre*, Te – *Teucrium chamaedrys*, Tp – *Thymus pulegioides*, Tr – *Trifolium repens*, Va – *Vicia angustifolia*, Vh – *Vicia hirsuta*, Vs – *Vicia tetrasperma*, Vt – *Vicia tenuifolia*.

Discussion

In most papers dealing with the species associations in the course of succession or on the disturbance gradient (Greig-Smith 1952, Kellman 1969, O'Connor and Aarssen 1987), attention is paid mainly to the frequency (total number standardized to the number of possible species pairs) of significant associations and its decrease in successionally late or undisturbed climax vegetation is usually found. In our study, we have not found a consistent trend in the frequency of significant associations, but in accordance with the

19

Table 3. Average values of the V coefficient for groups of species pairs belonging to particular life forms in the young field. Numbers of species in particular groups are given in parenthesis.

		annuals	biennials	perennial forbs
annuals	(19)	-0.017		
biennials	(2)	-0.018	-0.077	
perennial forbs	(5)	-0.010	0.004	-0.029

Monte Carlo test based on F-value: $P = 0.740$; Monte Carlo test based on difference between maximum and minimum: $P = 0.420$.

Table 4. Average values of the V coefficient for groups of species pairs belonging to particular life forms in the medium aged field. Numbers of species in particular groups are given in parenthesis.

		annuals	biennials	perennial grasses	perennial forbs
annuals	(7)	0.004			
biennials	(3)	-0.015	-0.031		
perennial grasses	(4)	-0.007	-0.030	-0.161	
perennial forbs	(10)	-0.014	-0.011	-0.024	-0.018

Monte Carlo test based on F-value: $P = 0.046$; Monte Carlo test based on difference between maximum and minimum: $P = 0.040$.

Table 5. Average values of the V coefficient for groups of species pairs belonging to particular life forms in the oldest field. Numbers of species in particular groups are given in parenthesis.

		annuals	perennial grasses	perennial forbs
annuals	(5)	0.036		
perennial grasses	(5)	-0.013	-0.048	
perennial forbs	(13)	-0.014	-0.022	-0.016

Monte Carlo test based on F-value: $P = 0.336$; Monte Carlo test based on difference between maximum and minimum: $P = 0.028$.

papers cited above, the minimum of associations was found in the mature forest. On the other hand, it seems that more information may be obtained, when the pattern of associations is analyzed with respect to life history of particular species.

In most cases, the number of significant associations is very low, close to the expected number of associations resulting from the Type I error (i.e. rejecting the null hypothesis when it is true). However, manipulative experiments with selective herbicides carried out on the same plots (Michálek, Buriánek and Hadincová 1990) supported the hypothesis about strong competition between species. Particularly, the exclusion or suppression of one group of species led to an increase of abundance in unaffected (less affected) species. For example, the exclusion of dominant grasses (*Festuca rupicola, Poa angustifolia*) on the oldest field resulted in rapid spread and dominance of the species *Fragaria viridis*. This is evidence for competitive suppression of *Fragaria*. The value of V is -0.12 for *Fragaria* and *Festuca* and 0.13 for *Fragaria* and *Poa* (both non-significant). The estimation of pairwise competition effect from manipulative experiments and their subsequent comparison with V values is impossible as the herbicides influenced various species to various extents. The experiments were not designed to detect and measure competition, nevertheless their results suggest that competition plays an important role in shaping the community, despite the fact that only few significant negative associations were detected. However, the differences in morphology of *Poa* and *Festuca* indicate that *Fragaria* might be more competitively suppressed by *Festuca*, which has more compact tussocks.

The low frequency of negative associations determined from qualitative data may be partially caused by the low frequencies of species — the expected number of common occurences is then close to zero and the negative association is undetectable. However, the number of positive associations and the number of associations determined from quantitative data is low, too. This supports the idea of importance of diffuse competition in plant communities (Wilson and Keddy 1986).

Neither Rogers (1983), nor we in the present study found a pronounced relationship between phenological similarity of species and their association. Nevertheless, we have found, particularly in older successional stages, non-random patterns in species associations with regard to species life histories. Annuals are usually positively associated, whereas between strong perennial competitors negative associations prevail.

Conclusions

Following conclusions can be drawn:
1. No consistent trend was found in the frequency of significant associations with the successional age of the plot.
2. The hypothesis that species with similar phenology compete more and consequently are spatially segregated is only weakly supported by the data.
3. In older successional stages, differences were found in associations of species with respect to their life history types.
4. The frequency of significant associations may be considered neither as a measure of biological organization, nor as a measure of competition in the community.

Acknowledgements. Thanks are due to Tomáš Havránek for critical reading of the first draft of the paper.

References

Goodall D.W.(1952): Some considerations in the use of point-quadrats for the analysis of vegetation.—Aust. J. Sci. Res. 5: 1–41.

Greig-Smith P. (1952): Ecological observations on degraded and secondary forest in Trinidad, British West Indies.—J. Ecol. 40: 316–330.

Hurlbert S.H. (1969): A coefficient of interspecific association.—Ecology 50: 1–9.

Kellman M.C. (1969): Plant species interrelationships in a secondary succession in Coastal British Columbia.—Syesis 2: 201–212.

Kershaw K.A. (1973): Quantitative and dynamic plant ecology. 2nd ed. — Edward Arnold, London.

Kershaw K.A. and Looney J.H.H. (1985): Quantitative and dynamic plant ecology. 3rd ed. — Edward Arnold, London.

Michálek J., Buriánek V. and Hadincová V. (1990): Effect of herbicides. — In: Osbornová J., Kovářová M., Lepš J. and Prach K. [eds.], Succession in abandoned fields. Studies in Central Bohemia, Czechoslovakia. Kluwer Publ., Dordrecht, pp. 116-125.

O'Connor I. and Aarssen L.W. (1987): Species association patterns in abandoned sand quarries.—Vegetatio 73: 101–109.

Osbornová J., Kovářová M., Lepš J. and Prach K. [eds.] (1990): Succession in abandoned fields. Studies in Central Bohemia, Czechoslovakia. Kluwer Publ., Dordrecht, 168 pp.

Rogers R.S. (1983): Small-area coexistence of vernal herbs: does functional similarity of plants matter?—Am. Nat. 121: 835–850.

Wilson S.D. and Keddy P.A. (1986): Measuring diffuse competition along an environmental gradient: results from a shoreline plant community.—Am. Nat. 127: 862–869.

Krahulec F., Agnew A.D.Q., Agnew S. & Willems J.H. [eds.]: *Spatial processes in plant communities.* pp. 23-30.

The effect of grazing on the structure of a sandy grassland

Gábor Matus[1] and Béla Tóthmérész[2]

1) *Botanical Department of Kossuth L. University, Debrecen, P.O.Box 14, H-4010, Hungary*
2) *Ecological Department of Kossuth L. University, Debrecen, P.O.Box 14, H-4010, Hungary*

Keywords: Association analysis, Species associations, Guild, Grazing, Community perturbation, Diversity, Principal coordinate analysis, Plexus graph.

Abstract. The structure of a grassland community (*Cynodonto-Festucetum pseudovinae*) was studied in East Hungary using transects. A cattle-grazed part of the grassland and a non-grazed one was compared in order to explore the effect of moderate cattle grazing on community structure.

Schluter's test of association suggests a positive net association of species. Deviations from independence are significant in both cases. A combination of principal coordinate analysis and plexus diagram method indicated two coalitions (guilds) of species. Species within coalitions are positively associated, but these groups are separated by negative associations. Both the number of positive and negative associations were smaller in the grazed case. The number of species, however, was little higher in the cattle grazed case.

Introduction

Some ecologists profess that herbivory plays a central role in the organization of plant communities. In some cases grazing can be a part of the ecosystem; i.e. we know such vegetation which can be controlled or even created by grazing. In other cases grazing means only a perturbation of the plant community. Grazing may alter the competitive balance between the species of the community, especially between the grazed plants and other species. Above ground grazing usually results in loss of below ground material as well, which influences root competition and access to mineral resources and water.

We were out to examine the effect of moderate cattle grazing on the structure of a sandy grassland community (*Cynodonto-Festucetum pseudovinae*) in Hungary. In this case a moderate cattle grazing means a perturbation because grazing is not an inherent part of the ecosystem here. We wanted to quantify the amount of changing in vegetation composition resulting from grazing; and we were especially interested in the changing of pairwise interactions between the species.

23

Association analysis is especially useful to gain insight the working mechanisms of plant communities. We applied this method in our study; the result of association analysis cannot be displayed objectively (i.e. the location of the species in the plexus graph is arbitrary), thus we combined the method with the principal coordinate analysis which is an especially useful method displaying "objectively" complex data structures in two (or more) dimensions. Until this time it was not a standard technique in association analysis.

Study area

The area of Daru-hegyek is situated in the Great Hungarian Plain 30 kilometers east of Debrecen, Hajdú-Bihar County. The area is characterized by parabola shaped sand-hills consisting of pleistocene wind blown sand. The average length of hills is 500–900 m, ranging in height from 3–10 m. The humus and lime content of the soil, derived from sand, are very low.

Two main plant communities can be found: a calcifuge (calciphobic) perennial sandy grassland (*Festuco vaginatae-Corynephoretum*) on steep slopes (constant: *Festuca vaginata, Corynephorus canescens, Euphorbia cyparissias*; frequent: *Alyssum montanum* ssp. *gmelini, Carex stenophylla, Potentilla arenaria, Thymus glabrescens, Anthemis ruthenica, Centaurea pseudarenaria*) and a perennial sandy pasture (*Cynodonto-Festucetum pseudovinae*) on hilltops and gentle slopes (constant: *Festuca pseudovina, Carex stenophylla, C. supina, Cerastium semidecandrum, Potentilla arenaria, Eryngium campestre*; frequent: *Cynodon dactylon, Kochia laniflora, Pulsatilla pratensis* ssp. *hungarica*) (Soó 1939).

At the foot of hills various hydrophilous communities can be found (*Caricetum appropinquatae, Carici-Menyanthemum, Carici-Calamagrostietum neglectae, Cirsio cani-Festucetum pratensis, Calamagrostio-Salicetum cinereae, Molinietum coeruleae*). The area is under protection because of some rare plant species living there (*Pulsatilla pratensis* ssp. *hungarica, Onosma arenaria, Iris humilis* ssp. *arenaria*). There are considerable differences in the intensity of cattle grazing between certain parts of the area.

Methods

Sampling

The sampling area was situated on a north-west slope of one of the typical sand-hills. Two transects were staked out in a *Cynodonto-Festucetum pseudovinae* community. One of them was in an ungrazed patch the other was in a patch grazed by cattle to a small extent. The transects were 2m in breadth and 16m in length from the top to the bottom of the hill, and were divided into 800 quadrats of 20 by 20 cm size. The number of individuals and shoots of higher plants was noted in each quadrat. Field investigations were made in July, 1988 (see Table 1).

Schluter's test of association

Schluter (1984) published a method to test the significance of associations among S species taken simultaneously (S is the number of species in the studied community; $S > 2$). The method is based on a new approach using a variance ratio derived from a null association model to test simultaneously for significant associations.

24

Table 1. A list of species detected in the transects. Nomenclature follows Soó and Kárpáti (1968). T_1 = transect in the ungrazed patch; T_2 = transect in the grazed patch (moderate cattle grazing). Both of the sampling areas covered $16 \times 2m$, the quadrat size was $20 \times 20cm$.

Species	Code	Frequency		Species	Code	Frequency	
		T_1	T_2			T_1	T_2
Achillea collina	ACM	-	131	*Koeleria glauca*	KOE	9	2
Achillea setacea	ACH	333	1432	*Kochia laniflora*	KOL	1897	3958
Agropyron repens	AGP	-	1	*Luzula campestris*	LUZ	60	152
Agrostis alba	AGR	366	34	*Plantago lanceolata*	PLA	60	152
Ambrosia elatior	AMB	1	2	*Potentilla arenaria*	POT	367	537
Anchusa officinalis	ANO	-	2	*Potentilla argentea*	POA	1	60
Anthemis ruthenica	ANT	274	251	*Rumex acetosa*	RUA	1	2
Asperula cynanchica	ASP	6	-	*Rumex acetosella*	RUM	673	867
Berteroa incana	BER	-	1	*Salvia pratensis*	SAL	1	1
Carex stenophylla	CAR	1998	4493	*Sedum sexangulare*	SEX	39	87
Crepis rhoedifolia	CRE	1	10	*Silene otites*	SIO	2	-
Cynodon dactylon	GRA	936	1766	*Teucrium chamaedrys*	TEU	-	993
Erigeron canadensis	ERI	15	1	*Thymus degenianus*	THY	896	362
Erodium cicutarium	ERO	4	11	*Trifolium arvense*	TRA	8	1
Eryngium campestre	ERY	9	16	*Trifolium campestre*	TRI	310	39
Erysimum diffusum	ERM	2	63	*Tunica prolifera*	TUN	1	10
Euphorbia cyparissias	EUC	216	155	*Verbascum phoeniceum*	VER	61	40
Equisetum ramosissimum	EQU	87	22	*Veronica austriaca*	VEA	13	18
Festuca pseudovina	FEE	2971	1178	*Veronica verna*	VEV	1	-
Galium verum	GAV	-	42				

p_i denotes the relative frequency of occurrence of species i in the quadrat samples, i.e. $p_i = n_i/N$, where n_i is the number of quadrats where the species was present and N is the total number of quadrats.

The total sample variance is calculated as

$$\text{Var}_{\text{sample}} = \sum_{i=1}^{S} p_i(1 - p_i),$$

where S is the total number of species in the samples.

The variance in total species number is estimated by

$$\text{Var}_{\text{species}} = \frac{\sum_{j=1}^{N}(T_j - t)^2}{N}$$

$$t = \frac{\sum_{j=1}^{N} T_j}{N},$$

where T_j is the number of species in the j-th sample and t is the mean number of species per sample. The VR variance ratio is defined by

$$VR = \frac{\text{Var}_{\text{species}}}{\text{Var}_{\text{sample}}}.$$

VR is an index of species association. The expected value of VR under the null hypothesis of independence is exactly 1. $VR > 1$ suggests that the species exhibit positive association. $VR < 1$ suggests a negative net association.

A W statistic which is defined by

$$W = N \times VR$$

may be used to test whether deviations from independence are significant. Critical values for rejecting the null hypothesis may be obtained from a table of critical values for the chi-square distribution. Under the hypothesis of no association, there is a 90% probability that W should lie between the limits

$$\chi^2_{0.05,N} < W < \chi^2_{0.95,N}.$$

It is evident that there might be situations where some species occur positively among themselves, and negatively with other species, so the Schluter's variance ratio test detects no associations.

The above described method is based on presence-absence data of species; however, Schluter extended his method to use density data of species. McCulloch (1985) examines and explaines the relation of Schluter's test and some standard statistical tests of association analysis.

Association analysis

The chi-square test of independence is a broadly used method for testing independence in a 2×2 table. The proper statistic is defined as follows:

$$\chi^2 = \frac{(ad - bc)^2 N}{(a + b)(c + d)(a + c)(b + d)},$$

where $N = a+b+c+d$, i.e. the total number of quadrats. In practice, the Yates' correction for continuity is used to eliminate sampling error. This is carried out by using the following expression:

$$\chi^2_{\text{corrected}} = \frac{(|ad - bc| - N/2)^2 N}{(a + b)(c + d)(a + c)(b + d)}.$$

26

The calculated $\chi^2_{\text{corrected}}$ should be compared with χ^2 for one degree of freedom. The Yates' correction tend to be overcompensate the sampling error (Greig-Smith 1983). Sokal and Rohlf (1981) prefer the G-test statistic to test independence.

Principal coordinate analysis

Principal coordinate analysis (PCoA) has also been referred to as a metric multidimensional scaling, or sometimes classical multidimensional scaling. PCoA takes an objects by objects distance matrix as input and produces coordinate values as output. This method is very powerful in such cases where we estimate the distances of objects in a multidimensional space by a rather complicated procedure and we would like to produce a relatively simple representation of objects in a low dimensional (frequently 2-dimensional) space.

Let assume that we have a $D = (d_{ij})$ distance matrix; $i, j = 1, ..., n$. We have to construct a matrix A from the distance using the following formula:

$$a_{ij} = \frac{-(d_{ij}^2 - d_i^2 - d_j^2 - d^2)}{2},$$

where

$$d_i^2 = \sum_{j=1}^{n} d_{ij}^2 / n^2,$$

$$d_j^2 = \sum_{i=1}^{n} d_{ij}^2 / n^2,$$

$$d^2 = \sum_{i=1}^{n} \sum_{j=1}^{n} d_{ij}^2 / n^2.$$

If matrix A is positive, symmetric and semidefinite, it will have rank $p \leq n$, so we may derive p non-zero eigenvalues, $\lambda_1, \lambda_2, ..., \lambda_p \neq 0$, with corresponding eigenvectors, $u_1, u_2, ..., u_p$. A coordinate matrix

$$C = \begin{pmatrix} c_{11} & \cdots & c_{1p} \\ c_{21} & & c_{2p} \\ . & & \\ . & & \\ c_{n1} & \cdots & c_{np} \end{pmatrix}$$

which is the result of PCoA is constructed by the rescaled eigenvectors

$$c_{ij} = \sqrt{\lambda_i} u_{ij}, i + 1, ..., n; j = 1, ..., p.$$

The method is similar to the principal component analysis. As a result of it we have a series of orthogonal axes which indicate the inherent dimensionality and which may be used for plotting the objects studied (Mardia $et\ al.$ 1979).

Results and Discussion

The number of species in the grazed area was 36 compared to 33 in the ungrazed one. The species composition was very similar. The similarity was 76.9% measured by the Jaccard similarity index and 86.9% measured by the Sörensen index. The similarity was reduced if we examined not only the species composition but the number of individuals as well. The similarity was 59.2% measured by the Czekanowski index (similarity version of Bray-Curtis dissimilarity) and 31.5% measured by the Sigleo index.

The number of rare species was approximately equal in both cases; there were 13 rare species in the ungrazed case and 12 in the grazed one. (A species was taken to be rare if the number of individuals was below 10.)

The Schluter's test of associations based on presence-absence data of species suggested a positive net association in both cases. Deviations from independence were significant; the probability level was 99.5%.

The association between the species was examined at the $p < 0.05$ significance level by chi-square statistic. The Yates' correction for continuity was used. We omitted the associations where more than 80% of the sampling units were 'empty', i.e. none of the studied species was present or only extremely rare. This omission decreases the risk of spurious correlation.

In the ungrazed case species *Achillea setacea, Agrostis alba, Eryngium campestre, Euphorbia cyparissias, Luzula campestris, Plantago lanceolata, Sedum sexangulare, Thymus degenianus, Trifolium campestre* and *Verbascum phoeniceum* were positively associated as were species *Anthemis ruthenica, Kochia laniflora* and *Rumex acetosella*, but these 2 groups were separated by negative associations. In the grazed case species *Achillea setacea, Euphorbia cyparissias, Festuca pseudovina, Potentilla anserina, Sedum sexangulare, Teucrium chamaedrys, Thymus degenianus* and *Trifolium campestre* were positively associated in the first coalition, species *Anthemis ruthenica* and *Kochia laniflora* were positively associated in the second one, but these coalitions were separated by negative associations.

Both the number of negative and positive associations were decreased in the grazed case. In the ungrazed case there were 15 positive and 20 negative associations, while in the grazed case there were 7 positive and 13 negative associations.

Displaying the results of association analysis we used a combination of principal coordinate analysis (PCoA) and plexus graph method. In the first step we eliminated the rare species. The distances between the species were calculated by the Euclidean distances with oblique coordinates on the basis of their occurrence (Orlóci 1978). The PCoA took this distance matrix as an input and produced coordinate values of species in the PCoA space as an input (Gordon 1981, Legendre and Legendre 1983). Plotting the results in a two dimensional space we joined the positively associated species by solid lines and negatively associated ones by dashed lines. Coalitions are well-demonstrated by the figures: the positively associated species are aggregated into two compact groups, and these groups, individuals of which are negatively associated, are definitely segregated (see Figure 1 and Figure 2).

The proportion of significant negative and positive interspecific associations differs between the two stands. Our results suggest that a low level of grazing increases the number of species and decreases the number of positive and negative associations, whilst

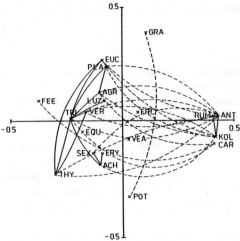

Figure 1. Plexus diagram for 20 herbaceous species of an unperturbed grassland community. The species are arranged in a hypothetical two-dimensional space by Principal Coordinate Analysis. Associations are analysed by the standard chi-square procedure. Species ANT, KOL and RUM are positively associated (solid lines) as do ACH, AGR, ERY, EUC, LUZ, PLA, SEX, THY, TRI and VER, but these groups are separated by negative associations (dashed lines). (Abbreviations are in Table 1.)

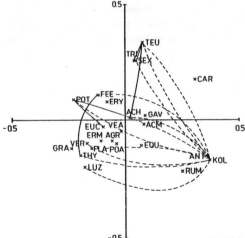

Figure 2. Plexus diagram for 24 herbaceous species of a perturbed (moderate cattle grazing) grassland community. Species ANT and KOL are positively associated (solid lines) as do ACH, EUC, FEE, POT, SEX, TEU, THY and TRI, but these groups are separated by negative associations (dashed lines). (Abbreviations are in Table 1.)

the species composition changes only slightly; the changes expressed in the number of individuals are also not high, although somewhat greater than reflected by the species

29

composition.

Acknowledgements. The authors are grateful to Prof. I. Précsényi for his help and useful advice.

References

Gordon A.D. (1981): Classification. Methods for the exploratory analysis of multivariate data.— Chapman and Hall, New York.

Greig-Smith P. (1983): Quantitative Plant Ecology. 3rd ed.—Blackwell Sci. Publ., Oxford.

Kershaw K.A. and Looney J.H. (1985): Quantitative and dynamic plant ecology. 3rd ed.— Edward Arnold.

Legendre L. and Legendre P. (1983): Numerical ecology.—Elsevier Sci. Publ. Co., Amsterdam.

Ludwig J.A. and Reynolds J.F. (1988): Statistical ecology.—Wiley, New York.

MacIntosh R.P. (1978): Matrix and plexus techniques.—In: R.H. Whittaker [ed.], Ordination of plant communities. Dr. W.Junk bv Publishers The Hague, pp. 151–184.

Mardia K.V., Kent J.T. and Bibby J.M. (1979): Multivariate analysis.—Academic Press, London.

McCulloch C.E. (1985): Variance tests for species association.—Ecology 66: 1676–1681.

Orlóci L. (1978): Multivariate analysis in vegetation research.—W. Junk, The Hague.

Schluter D. (1984): A variance test for detecting species associations.—Ecology 65: 998–1005.

Sokal R.R. and Rohlf F.J. (1981): Biometry. The principles and practice of statistics in biological research. 2nd ed.—W.H. Freeman and Co., New York.

Soó R. (1939): A Nyírség vegetációja III. A Nyírség pusztai növényszövetkezetei. [Vegetation of the Nyírség III. Grassland communities of the Nyírség.]—Magyar Tudományos Akadémia Matematikai és Természettudományi Értesítöje, LVIII., Budapest.

Soó R. and Kárpáti Z. (1968): Magyar flóra. [Flora of Hungaria] Tankönyvkiadó, Budapest.

Whittaker R.H. [ed.](1978): Ordination of Plant Communities.—Dr. W. Junk Publishers, The Hague, Boston.

Krahulec F., Agnew A.D.Q., Agnew S. & S. & Willems J.H. [eds.]: *Spatial processes in plant communities*. pp. 31–47.

Spatial processes in developing plant communities: pattern formation detected using information theory

Sándor Bartha

Research Institute of Ecology and Botany, Hungarian Academy of Sciences, H-2163 Vácrátót, Hungary

Keywords: Characteristic areas, Diversity, Mine waste, Spatial dependence, Species coalitions, Succession

Abstract. Development of topographical pattern during primary succession of plant assemblages was studied on dumps from open-cast coal mining in Hungary. Micromaps of the distribution of 11 dominant species were recorded in six plots of different age. The rate of succession was very high in the early spontaneous revegetation of spoil banks. Considering the temporal patterns we distinguished three groups of dominant species. Primary colonizers (coalition I): *Lactuca serriola*, *Matricaria inodora*, *Atriplex nitens* were followed by ruderal annuals and biennials (coalition II): *Bromus japonicus*, *Medicago lupulina*, *Melilotus officinalis* and coalition III (mostly perennials): *Agropyron repens*, *Lathyrus tuberosus*, *Rubus caesius* and *Picris hieracioides*. *Tussilago farfara* was among the primary colonizers but it could maintain its dominance into the later stages with its well developed vegetative clones. A family of information theory models developed by Juhász-Nagy proved to be powerful tools for describing complex spatial relations and their changes. Considerable changes of topographical patterns of species were detected during the first 10 years of primary succession. The areas where characteristic functions reached their maxima were different in each year. Regarding maximum values of florula diversity and associatum a generally increasing trend was found with time. However the trend was not monotonous: it had a sharp breaking point between 3 and 4 years at the transition between coalitions I and II. The transition between coalitions II and III was gradual in time. Considering spatial scales we found the highest level of patchiness at the middle phase of the establishment of coalition I and at the transition between coalitions II and III.

Introduction

Some recent models of succession focussed on mechanisms producing the sequence of species in a particular site after some perturbation (Connell and Slatyer 1977). Characteristics like change of species composition in time or the dynamic status of species in the sequence were successfully related to population biological attributes (Bazzaz 1979, Noble and Slatyer 1980, Peet and Christensen 1980, Pickett 1982, Brown and Southwood 1987). However, increasing evidence suggests that local heterogeneities in environmental factors and in species availability generate spatially variable processes (Glenn-Lewin 1980, Game *et al.* 1982, Numata 1982, Humphrey 1984, Gibson *et al.* 1985, Russel and LaRoi 1986, Palmer 1987, Wood and del Moral 1987, Lepš 1987, Prach 1987, and Cowling and Pierce 1988) and the created patchiness of vegetation has important consequences for species interactions, rate of change and successional pathways (van Andel and Nelissen 1981, Czárán 1984, Pickett and White 1985, Prach 1987, Peart 1989a, 1989b, 1989c, Czárán and Bartha 1989). Despite its obvious importance, relatively few studies express generalizations about the development of topographical pattern in succession (Greig-Smith 1964, Juhász-Nagy 1980). Information theory methods used in this paper have some special advantages in describing spatial dependence of plant populations (Juhász-Nagy 1976, 1980, 1984, Juhász-Nagy and Podani 1983). The results presented here are parts of a larger research program investigating pattern-formation of plant assemblages in the early stages of succession on dumps from open-cast coal mining. When mining activity completely destroys large areas it creates very suitable conditions for studying rapid autogenic vegetation dynamics (Prach 1984, 1986a, 1986b and 1987).

Materials and Methods

Study site

The mined area is located in the northern part of Hungary, at the southern foot of the Mátra Mountains, at 47°16′N, 20°6′E, ca. 90 km NE of Budapest (Fig.1.). The mean annual temperature is 10.5°C and the average annual precipitation 582 mm (Oláh and Szegi 1979). The waste soil dumps consist mainly of different mixtures of Pannonian clay, Pannonian sand, secondary Andesitic tuff and Holocene sediment. The study sites were located on dumps consisting predominantly of Pannonian yellow clay. The main chemical and physical properties of Pannonian clay are: pH 7.5, $CaCO_3$ 15.4%, total organic matter 0.09%, total salts 0.02%; available nutrients (mg per 100/g soil): N 0.42, P_2O_5 27.4, K_2O 11.0, stichy point (water saturation percent) according to Arany as saturation percentage (SP): 68, sand:clay ratio 25.2:74.4. The general soil conditions (toxic compounds not found, pH is not low) provide rather good conditions for rapid recolonization of dumps by weed species (Szegi *et al.* 1988).

Sampling procedure

Micromaps of species distributions were made in six plots of age 1, 2, 3, 4, 7 and 10 years. To ensure the accurate estimation of species patterns we had to choose plots different in area because of the changing scale of patchiness. The size of the plots were 10 × 10 m for 3, 4 and 10 year old stands, 30 × 15 m for 1 year old, 35 × 25 m for 2 year old and 22 × 15 m for 7 year old stand. Square lattices were used for sampling, the size of the units being 20 × 20 cm. Presence/absence data were recorded in each unit, for each species. Sampling procedures were performed each year between 1984 and 1987, in late summer (i.e. the phenologically best developed period of weed communities). Only undisturbed, spontaneously revegetated areas were investigated. Permanent studies were unfortunately impossible here because of unpredictable mining and post mining activities destroying the area. In the recent study, only 11 dominant weed species were used to avoid the uncertainties and random effects caused by rare species. Random references were generated by computer, positioning individuals of species with the same frequencies as in

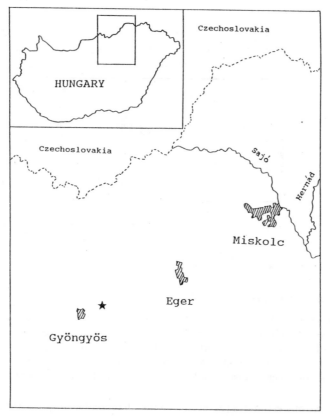

Fig.1. Geographical location of the study area in Hungary. Site marked by an asterisk.

the field. This method helps to distinguish between textural and structural influences on the functions calculated. From the basic (field and random reference) maps computerized sampling was then performed (Podani 1987) with increasing isodiametric sampling unit sizes (referred to in this article as spatial process). The number of quadrats sampled was 1000 in each step of analysis.

Data analyses

Information theory functions developed by Juhász-Nagy in the framework of a general diversity concept were used to describe spatial patterns. Since Shannon's intuitive invention of information theory, a number of mathematicians (Khinchin 1957, Kullback 1959, Rényi 1962) have developed a deeper and more useful foundation of the theory which is used below. Our terminology will follow their concepts. A more detailed comprehensive description of the concepts and models can be found in Juhász-Nagy (1984). For an introduction into these techniques see Juhász-Nagy and Podani (1983). (Some parts of this article will be cited here.) All the functions were calculated at increasing quadrat sizes (spatial process, Podani 1984).

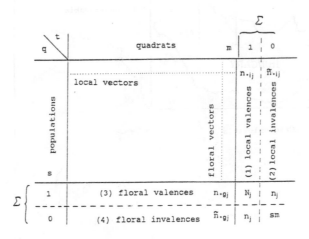

Fig. 2. Summary table of s species occurrences in m sampling units to give sm total cells.

1. Weighted marginal entropy functions

Fig 2. shows some notations for a binary coenological table (commonly used in phytosociological work). The marginals present the simple sums of presences (valences) and also the sums of absences (invalences). $n_{.ij}$ represents the sum of presences and $\tilde{n}_{.ij}$ the sum of absences for species i in m quadrats at quadrat size j and respectively $n_{.gj}$ represents the number of species present and $\tilde{n}_{.gj}$ the number of species absent in quadrat g at quadrat size j.

Weighted entropy estimates (according to the Shannon formula) for the four marginal distributions are:

(1) $N_j \hat{H}_j(V_q) = C_j - A_j$

(2) $N_j \hat{H}_j(V_t) = C_j - B_j$

(3) $n_j \hat{H}_j(v_q) = c_j - a_j$

(4) $n_j \hat{H}_j(v_t) = c_j - b_j$

where j refers to the quadrat size, q and t to the populations (local marginals) and quadrats (floral marginals) respectively. Capitals are used for valences and small letters for invalences. Further definitions (to make the calculations more easy):

$C_j = N_j \log N_j,$ $c_j = n_j \log n_j,$ where $N_j = \sum_{i=1}^{s} n_{.ij} = \sum_{g=1}^{m} n_{.gj}$

$A_j = \sum_{i=1}^{s} n_{.ij} \log n_{.ij}$ $a_j = \sum_{i=1}^{s} \tilde{n}_{.ij} \log \tilde{n}_{.ij}$ and $n_j = \sum_{i=1}^{s} \tilde{n}_{.ij} = \sum_{g=1}^{m} \tilde{n}_{.gj}$

$B_j = \sum_{q=1}^{m} n_{.qj} \log n_{.qj}$ $b_j = \sum_{q=1}^{m} n_{.gj} \log \tilde{n}_{.gj}$

(log means $\log_2$ in this paper, and numerical values will be given in Bits, i.e. properly weighted bits.)

The functions are all bounded by very simple inequalities:

(1) $N_j \log \bar{s}_j \leq N_j \hat{H}_j(V_q) \leq N_j \log s$

(2) $N_j \log \bar{m}_j \leq N_j \hat{H}_j(V_t) \leq N_j \log m$

(3) $n_j \log(s - \bar{s}_j) \leq n_j \hat{H}_j(v_q) \leq n_j \log s$

(4) $n_j \log(m - \bar{m}_j) \leq n_j \hat{H}_j(v_t) \leq n_j \log m$

34

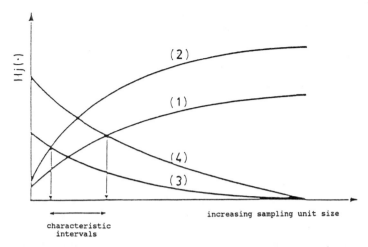

Fig. 3. A quartet of "species area curves" showing the values of the marginal entropy estimates in respect of valences (1 and 2) and invalences (3 and 4) with increasing sampling unit size.

where s is number of species, m is number of quadrats, $\bar{s}_j = N_j/m$ called local mean density and $\bar{m}_j = N_j/s$ called floral mean density. These marginal uncertainty estimates calculated for each step of spatial process follow monotonously increasing (1), (2) and monotonously decreasing (3), (4) patterns (Fig 3.). Entropy estimates of local marginals reflect the changes of commonness-rarity relations of species through the spatial process. These functions reach their extremum values (maximum for (1) and minimum for (3)) at larger quadrat size if there are more rare species in the community or more species with aggregated distribution. Entropy estimates of floral marginals detect the saturation of quadrats with species through the spatial process. If the number of gaps, the monodominant- or low diversity patches are increasing in the community these functions will approach their extremum values (maximum for (2) and minimum for (4)) slower (at larger quadrat size).

These functions might be the optimal substituents of the classic "species area curves" (Juhász-Nagy 1984) because their constants are very simple and clear, confronting the rather large set of species area functions with their "mystic" and uninterpretable constants. A further advantage of these functions is that they are consistent with a family of models describing community state in more detailed ways. The appropriate equalities of $N_j \hat{H}_j(V_t) = n_j \hat{H}_j(v_q)$ and $n_j \hat{H}_j(v_t) = N_j \hat{H}_j(V_q)$ define a very important interval of spatial scale (as a characteristic interval for a particular pattern) where (in the majority of cases) the maxima of other functions are all located (Juhász-Nagy 1984).

2. Functions for basic coenological phenomena

The following measures refer to the basic coenological phenomena (preference, diversity and resemblance) describing a particular stand of community (coenological state). Because of their additive nature they also define some bound characteristics of coenological states considered (in the sense of "averages" and "totals") but in a more detailed way.

35

Local distinctiveness. "Shannon's entropy function may be used to estimate the uncertainty of the event that a particular species occurs in a randomly chosen quadrat. If species i is present in $n_{.ij}$ of m sampling units, its entropy estimate is

$$m\hat{H}_j(i) = m\log m - (n_{.ij}\log n_{.ij} + \tilde{n}_{.ij}\log \tilde{n}_{.ij}).$$

Summation over all species will give the pooled entropy estimate,

$$m\hat{H}_j([L]) = \sum_{i=1}^{s} m\hat{H}_j(i)$$

which will be termed the local distinctiveness of species occurring in the study area." (Juhász-Nagy and Podani 1983, p. 132.) Local distinctiveness and local entropies reflect the preference of species. The local entropy of species i reaches its maximum when $n_{.ij} = m/2$ (the species is present in the half of the m sampling units). If species i "prefers" the study area it will be frequent and reaches the status of $n_{.ij} = m/2$ at relatively small quadrat sizes. Rare species will have their local entropy maxima at larger quadrat sizes. If species i has an aggregated distribution (prefers only particular patches in the study area) it will reaches maximum local entropy at a larger quadrat size than its randomly distributed counterpart. Local distinctiveness reaches its maximum where most of the component populations reach their maximum heterogeneity. (It is important to note that the species are all considered independently here.) The quadrat size where local distinctiveness has its maximum is one of the characteristic area points of communities called "compensatory area" (for more detailed explanation see Juhász-Nagy and Podani 1983).

Florula diversity. All communities have some kind of inside heterogeneity. The areas of particular species combinations and their spatial relations are essential in understanding the spatial behavior of communities, their "working mechanisms" (Watt 1947). It is rather hard to understand why these relations were and are neglected in the majority of works considering spatial processes (for example Greig-Smith 1964). "The concept of florula diversity is illustrated with two species, which is the simplest case. The joint entropy of species A and B is

$$m\hat{H}_j(A,B) = m\log m - a\log a - b\log b - c\log c - d\log d$$

Symbols a, b, c and d correspond to the notations of a 2×2 contingency table, $m = a + b + c + d$ is the number of quadrats. In this case florula diversity can be considered as the weighted estimate of Shannon's entropy for the set of possible combinations of two species in question. The generalization of $m\hat{H}_j(A,B)$ to s species is given by the formula

$$m\hat{H}_j(A,B,...,S) = m\log m - \sum_{k=1}^{\omega} f_k \log f_k$$

where f_k is the frequency of the kth species combination in the data set and $\omega = 2^s$ is the number of possible species combinations (different potencial floral vectors e.g. florulas)." (Juhász-Nagy and Podani 1983, p. 131.).

36

Maximum area of florula diversity (the quadrat size where this function reaches its maximum) is also characteristic of a community.

Associatum. Associatum is a resemblance measure for estimating total interlocal dependence of species. "In information theory the association between two species may be expressed in terms of their mutual information:

$$M\hat{I}_j(A, B) = m \log m + a \log a + b \log b + c \log c + d \log d - (a + c) \log(a + c) -$$

$$-(b + d) \log(b + d) - (a + b) \log(a + b) - (c + d) \log(c + d)$$

In general, the mutual information of s species is termed the contingency information involving an s-dimensional contingency table λ containing 2^s cells. This quantity may be simply derived by subtraction of

$$m\hat{H}_j([L]) - m\hat{H}_j(A, B, ..., S) = m\hat{I}_j(\lambda)$$

To distinguish between the two species case for which association has been used and the general situation, $m\hat{I}_j(\lambda)$ will be termed the associatum of s species." (Juhász-Nagy and Podani 1983, p. 132.) The quadrat size where associatum has its maxima defines the scale at which most significant species relationships are detectable.

Relative measures. The functions introduced above are strongly influenced by their constants m and s. It may be useful to define relative measures where actual values are related to their possible maxima. For example a simple measure called relative associatum may be a useful tool for comparing the degree of spatial dependence between stands with different s. Relative associatum is defined by associatum/florula diversity (Florula diversity is considered as the possible maximum for associatum.)

3. Characteristic ordering

Fig. 4A presents realistic examples of the information theory functions for the 7 years old stand. We can realize that local distinctiveness, florula diversity and associatum change considerably with increasing the quadrat size. The spatial scales (sampling unit sizes) where these functions reach their maxima differ between the field data and their random references. The fact that characteristic areas appear at larger spatial scales for the field data detects contagious patterns of individual species (local distinctiveness) and species combinations (florula diversity). The second small maximum of local distinctiveness and florula diversity for the random pattern refers some species with very low frequencies in the study area. Associatum shows two peaks for the field data while its values are about zero for the random reference. Fig. 4B shows the projections of maxima to the abcissa. Areas belonging to the maxima of particular functions characterize the community states well regarding spatial pattern. Fig. 4C depicts the behaviour of the marginal entropy functions in the spatial process for the 7 years old stand. These "species-area" functions approach their extremum values slower comparing to the random references. Fig.4D shows the characteristic intervals defined by the $N_j\hat{H}_j(V_t) = n_j\hat{H}_j(v_q)$ and $n_j\hat{H}_j(v_t) = N_j\hat{H}_j(V_q)$ equalities and the positions of characteristic area points within the intervals. This arrangement — called characteristic ordering — is very useful for short, simple descriptions of rather complex coenological states (Juhász-Nagy 1967).

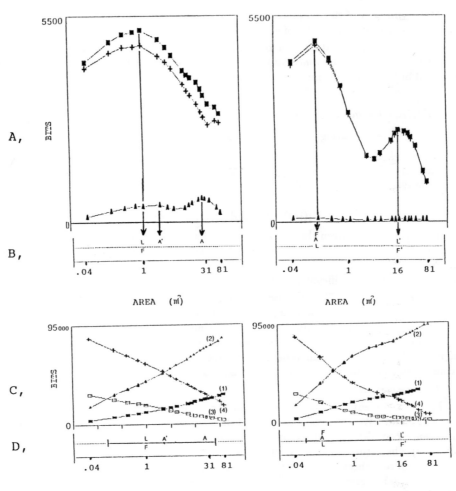

FIELD DATA   RANDOM REFERENCE

Fig. 4. Functions for basic coenological phenomena (an example with the 7-year-old stand).

A. Changes of local distinctiveness (■), florula diversity (+), and associatum (▲) with increasing quadrat size.

B. Projections of maxima to the abcissa (characteristic maximum areas). L — maximum area of local distinctiveness; F — maximum area of florula diversity; A — maximum area of associatum. (Apostrophes denote secondary peaks if present.)

C. Marginal entropy estimates through the spatial process. (■) $C_j - A_j$ (1); (▲) $C_j - B_j$ (2); (□) $c_j - a_j$ (3); (+) $c_j - b_j$ (4).

D. Characteristic intervals with the characteristic maximum area points (characteristic ordering). ⊢────┤ characteristic intervals

38

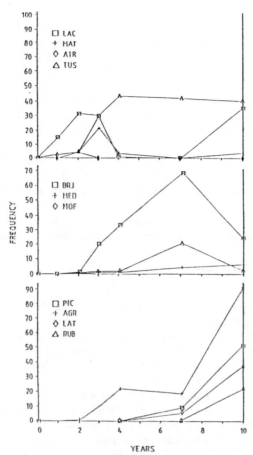

YEARS

Fig.5. Temporal changes of abundance for 11 dominant plant species. (Abbreviations: LAC *Lactuca serriola*; MAT *Matricaria inodora*; ATR *Atriplex nitens*; TUS *Tussilago farfara*; BRJ *Bromus japonicus*; MED *Medicago lupulina*; MOF *Melilotus officinalis*; PIC *Picris hieracioides*; AGR *Agropyron repens*; LAT *Lathyrus tuberosus*; RUB *Rubus caesius*.)

Results and Discussion

Changes of species abundances

Fig. 5 shows the temporal pattern of the 11 most important species in the first 10 years of the process. Years 1–3 are characterized by annual weed species [1] *Lactuca serriola* (LAC), *Matricaria inodora* (MAT) and *Atriplex nitens* (ATR) and the perennial *Tussilago farfara* (TUS). They can easily colonize the newly created surfaces from surrounding ruderal places because of the relatively good physical and chemical initial substratum

---

[1] Nomenclature: Soó R. and Jávorka S. (1951): A magyar növényvilág kézikönyve I-II. [Manual of the Hungarian flora] Akadémiai Kiadó, Budapest.

conditions and their large reproductive potential. The effects of increasing density, soil compaction, decreasing nutrient pools and water content make individuals of the dominant *Lactuca* and *Matricaria* stunted. *Tussilago farfara* with its deep rhizome system can maintain itself and becomes dominant in the 4th year. Several species important in the later stages are already present in the first two years and slowly occupy places in the form of small rather monodominant patches. The next group of species: the annual *Bromus japonicus* (BRJ), *Medicago lupulina* (MED) and the biennial *Melilotus officinalis* (MOF) with their intermediate invasive, dispersal and reproductive ability and intermediate competitive effort become dominant of the next stage ca. 4–8 years. They are gradually replaced by the biennial *Picris hieracioides* (PIC) and by clones of the strong competitor perennial *Agropyron repens* (AGR), *Lathyrus tuberosus* (LAT) and *Rubus caesius* (RUB). *Tussilago farfara* can still maintain its position probably by a competitive mass effect resulting from its well established vegetative clones. Field vole (*Microtus arvalis*) activity was observed from the 6th year of succession. The voles, devastating small areas, helped the pioneer species *Lactuca* and *Matricaria* re-establish themselves on relatively old sites.

Developing spatial patterns

1. Characteristic ordering

Characteristic intervals and points detect very strong changes of topographical patterns of species during the first 10 years of primary succession (Fig 6.). Because of the strong differences in characteristic areas, logarithmic scales were preferred in the abscissa. The characteristic interval in the first year of colonization is from 0.2 m^2 to 81 m^2, detecting a rather indeterminate state. It becomes shorter (from 1 m^2 to 81 m^2) for the second year and later gradually becomes even shorter and more compact (moves to the left side) resulting in the 10 year old stand having a length from ca. 0.04 m^2 to ca. 1 m^2. The only exception is the 7 year old stand with an expanded length moved to the right side of the scale caused by some rare species just appearing in the area like *Lathyrus* and *Rubus* and also by the increasing intensity of patchiness. The intervals calculated from the random references of field patterns show similar trends expressing the textural effects on the spatial patterns but they are moved to the left compared to the field cases and are shorter. These differences prove the existence of intensive contagious spatial pattern in each year of succession except in the first year, where the two intervals are almost the same, detecting random initial conditions.

Compensatory areas (L) (areas of the maxima of local distinctiveness) quantify the changing scales of patchiness. Comparing the field case with its random references they are exactly at the same positions in the first year (at 0.4 m^2) confirming the conclusions above. Random L moves to 1 m^2 in the second year detecting the textural effect of the appearance of new rare species (*Agropyron, Medicago* and *Melilotus*) while field L changes more (to 5.8 m^2) detecting the contagious pattern of the growing patches of individuals dispersed locally around their parents and also the effect of microtopography on the establishment of seedlings. In the 3rd years dominant species, *Lactuca, Matricaria* and *Tussilago* because of their good reproductive effort cover almost all the area, resulting in dense, rather random patterns (field L and random L are at the same, smaller scale of 0.2 m^2) while another maximum of field L, marked by L', appears at the same scale as in the second year. This is

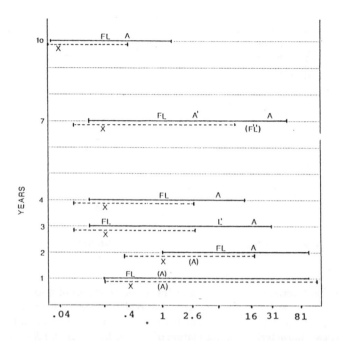

increasing sampling unit size

AREA (m²)

Fig 6. Characteristic ordering during the early primary succession.

├──────┤ characteristic intervals for field data

├ - - - - ┤ characteristic intervals for random reference

X = F,L (or F,L and A) at the same position.

(Apostrophes denote secondary peaks if present, other notations are the same as on Fig.3.B.)

Table 1. The number of species found in the particular years (from the 11 dominants)

years	1 2 3 4 7 10
number of species	4 8 8 7 10 10

probably caused by the disappearing *Atriplex* and the very slowly growing initial patches of *Agropyron, Medicago* and *Melilotus*. In the 4th year when early abundant species (*Lactuca, Matricaria*) almost disappeared while *Agropyron, Medicago* and *Melilotus* increased their clones, well expressed contagious pattern is detected again. The average scale of patchiness remains the same for the 7th year (0.2 m² for random L and 1 m² for field L) despite changes in the number of species (Table 1.) and the substitution of their dominance relations (*Bromus, Tussilago* and *Melilotus* are predominant now). However some changes

41

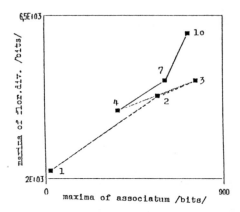

Fig. 7. Transformations of coenological states during the early primary succession (the numbers indicate years).

were detected in characteristic intervals and in the absolute values of maxima (referred to later in the text). The overdominance of *Agropyron repens* in the 10 year old stand alters the characteristic points abruptly to such small scales as 0.04 m² for random L and 0.2 m² for field L. Biological characteristic of population of *Picris, Lathyrus* and *Rubus* allow them to coexist fairly well with *Agropyron*. They can be randomly scattered, or creeping and climbing rather freely in the stand of *Agropyron*. The local effect of field voles (resulting in the re-establishment of *Lactuca* and *Matricaria*) and also the small patches of *Medicago* and disappearing *Melilotus* result in the small level of contagious pattern detected. The maximum areas of florula diversity (F) appeared at the same positions as maximum areas of corresponding local distinctiveness (L). It suggests a very low level of complexity without higher level species multiplets (multi-species coalitions) and makes us interpret spatial dependence relations very carefully. Areas of associatum maxima (A) of field data follow the relation: A > F = L. This means that maximum spatial dependence focussed on the subordinated and rare species. Common species predominant in the given years probably express more or less random pattern. Maximum areas of associatum in the random cases are at the same position as random F and random L. The exceptions (years 1 and 2) reflect the mutual effect of textural constraints and the imperfect random number generator used.

2. Phase space representations

Considering other characteristics Fig. 7 shows a special representation of absolute values of florula diversity and associatum. This kind of phase space representation depicts mutual behavior of diversity and resemblance generating a space where each point represents a possible community state. Trajectories detected follow a general increasing trend with time: diversity of realized species combinations and also the degree of spatial dependence increase in time. But the trend detected is not monotonous: it has a breaking point between 3 and 4 years. According to their population biological characters and our field experience three groups of species (coalitions) can be distinguished: early pioneers (mainly

42

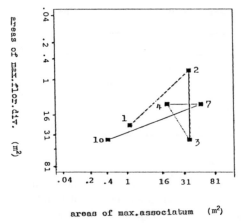

Fig. 8. Transformations of characteristic areas during the early primary succession (the numbers indicate years).

annuals): *Lactuca, Matricaria, Atriplex* and *Tussilago*; middle successional ruderals (annuals and biennials): *Bromus, Medicago, Melilotus*; and late successional species (mainly perennials): *Agropyron, Picris, Lathyrus* and *Rubus*. The singular point (breaking point) detected is at the transition between coalitions I and II. Similar behaviour between coalitions II and III was not found probably because of the slow, gradual transition between them. Further data from years 5, 6, 8 and 9 should be collected to clarify this problem. Similar mutual representations for the spatial scales (characteristic areas) are depicted in Fig. 8. The complex patterns can be simplified if we distinguish two stages: years 1–3 and years 4–10. In this case special "hairpin" forms are detected expressing a trend which starts with sporadic individuals appearing at relatively small or intermediate quadrat size. This is followed by intensive patchiness at a characteristically increasing scale in the middle stage of establishment of populations. The well established dominants develop a rather homogeneous well mixed pattern at smaller quadrat sizes. The directions of "hairpins" change with time. In the first (1 to 3 year) phase, at a smaller degree of spatial dependence, the main changes are expressed in the transformations of the characteristic areas of florula diversity. In the later phase (4 to 10 years), which is rather saturated in the sense of realized species combinations, associatum is responsible for the main changes in spatial scales.

3. Relative measures

Relative associatum expresses rather small values in the early (1 to 10 years old) period of succession investigated (Table 2.). Although only 10–30% of possible maxima are realized, these values are considerable higher than the corresponding random references (except in the first year). Their trends are not monotonous in accordance with the replacement of phases (coalitions) and the accompanying abrupt changes in community patterns.

43

Tab. 2. Relative associatum (associatum/florula diversity) during the early primary succession.

years	1	2	3	4	7	10
relative associatum field data	1.1	15.1	32.2	12.6	20.7	14.4
relative associatum random reference	0.2	2.3	2.1	1.7	2.0	5.7

Trends in pattern formation

Many authors stressed the effect of quadrat size on the results of vegetation analysis (for an excellent review see Podani 1984). Scale dependent trends of vegetation processes were found for example by Matthews (1979), de Pablo et al. (1982), Juhász-Nagy and Podani (1983), Hogeweg et al. (1985), O'Connor and Aarssen (1987), and Lepš and Štursa (1989). Beeftink (1987) suggested that "scale" and "detail" are variables in the study of dynamics in vegetation, where "scale" means the total size of area investigated and "detail" refers to the level of discrimination (sampling unit size). The information theory methods developed by Juhász-Nagy (1984) proved to be powerful tools for describing complex spatial relations and their changes. The considerable rearrangements of the characteristic areas during succession support the view that the detection of vegetation processes is inherently scale-dependent. We have to use a series of increasing quadrat sizes in order to obtain correct comparable results. In accordance with Juhász-Nagy (1980) and Juhász-Nagy and Podani (1983) we found a general decrease of characteristic areas and general increase of absolute maxima of characteristic functions during succession but the trajectories detected were more complicated with non-monotonous trends. Greig-Smith (1964) described the hypothetical trend of pattern development of particular species during succession. He suggested that this development is determined by the establishment and growth of patches of population. As patches enlarge, the scale of pattern will increase to a maximum at its optimum in the succession. The nucleation model of Yarranton and Morrison (1974) suggested that the changes of spatial pattern described by Greig-Smith for successional processes are repeated during each stage of the process. Morrison and Yarranton (1973) showed for sand dune succession that heterogeneity is more expressed at the transition between stages. In our cases multi-species patterns for an assemblage of dominants were detected producing a more complicated situation by the coincidence of individual trends. The multi-species trend, superposed from individual species, depends on the species-packing relationships on the successional gradient. Thus in our case we have detected three coalitions considering population biological characters and only two stages considering trajectories of pattern development. (Coalitions II and III were strongly overlapping in time but relatively well distinguished from coalition I.) The problem that maximum heterogeneity does not always coincide with stage-to-stage transitions is also related to this species-packing effect.

Acknowledgements. The author is grateful to A. Hilbert, G. Fekete, A. Borhidi and T. Czárán for their help in field work and to P. Juhász-Nagy for discussions. The constructive critics and suggestions by A.D.Q. Agnew and an anonymous reviewer are also acknowledged.

References

van Andel J. and Nelissen H. J. M. (1981): An experimental approach to the study of species interference in a patchy vegetation.—Vegetatio 45: 155–163.

Bazzaz F. A. (1979): The physiological ecology of plant succession.—Ann. Rev. Ecol. Syst. 10: 351–371.

Beeftink W. G. (1987): Scale and detail as variables in the study of dynamics in vegetation.— Abstracts of the General lectures, Symposium Papers and Posters p. 364. ; XIV. International Botanical Congress, Berlin (West), Germany, 24 July to 1 August.

Brown V. K. and Southwood T. R. E. (1987): Secondary succession: patterns and strategies.—In: Gray A. J., Crawley M. J. and Edwards P. J. [eds.], Colonization, succession and stability.— Blackwell Sci. Publ., Oxford.

Czárán T. (1984): A simulation model for generating patterns of sessile populations.—Abstracta Botanica 8: 1–13.

Czárán T. and Bartha S. (1989) The effect of spatial pattern on community dynamics: a comparison of simulated and field data.—Vegetatio 83: 229–239.

Connell J. H. and Slatyer R. O. (1977): Mechanisms of succession in natural communities and their role in community stability and organization.—Amer. Nat. 111: 1119–1144.

Cowling R. M. and Pierce S. M. (1988): Secondary succession in coastal dune fynbos: variation due to site and disturbance.— Vegetatio 76: 131–139.

Game M., Carrel J. E. and Hotrabhavandra T. (1982): Patch dynamics of plant succession on abandoned surface coal mines: a case history approach.—J. Ecol. 70: 707–720.

Gibson D. J., Johnson F. L. and Risser P. G. (1985): Revegetation of unreclaimed coal strip mines in Oklahoma. II. Plant communities.—Reclamation and Revegetation Research 4: 31–47.

Glenn-Lewin D. C. (1980): The individualistic nature of plant community development.— Vegetatio 43: 141–146.

Greig-Smith P. (1964): Quantitative plant ecology.—2nd. ed. Butterworth, London.

Hogeweg P., Hesper B., van Schaik C. P. and Beeftink W.G. (1985): Patterns in vegetation succession, an ecomorphological study.—In: White J. [ed.], The population structure of vegetation.—Dr.W.Junk Publ., Dordrecht, p. 637–666.

Humphrey L. D. (1984): Patterns and mechanisms of plant succession after fire on *Artemisia*-grass site in southeastern Idaho.—Vegetatio 57: 91–101.

Juhász-Nagy P. (1967): On some 'characteristic area' of plant community stands.—Proc. Colloq. Inf. Theory, Bolyai Math. Soc., Debrecen pp. 269–282.

Juhász-Nagy P. (1976): Spatial dependence of plant populations. Part 1. Equivalence analysis (An outline of a new model).—Acta Bot. Acad. Sci. Hung. 22: 61–78.

Juhász-Nagy P. (1980): A cönológia koegzisztenciális szerkezeteinek modellezése.—Doctoral Thesis [Manuscript, in Hungarian], Budapest.

Juhász-Nagy, P. (1984): Spatial dependence of plant populations. Part 2. A family of new models.—Acta Bot. Acad. Sci. Hung. 30: 363–402.

Juhász-Nagy P. and Podani J. (1983): Information theory methods for the study of spatial processes and succession.—Vegetatio 51: 129–140.

Khinchin A. I. (1957): Mathematical foundations of information theory.—Dover, New York.

Kullback S. (1959): Information theory and statistics.— Wiley, New York.

Lepš J. (1987): Vegetation dynamics in early old field succession: a quantitative approach.— Vegetatio 72: 95–102.

Lepš J. and Štursa J. (1989): Species area curve, life history strategies, and succession: a field test of relationships.—Vegetatio 83: 249–257.

Matthews J. A. (1979): A study of the variability of some successional and climax plant assemblage-types using multiple discriminant analysis.—J. Ecol. 67: 255–271.

Morrison R. G. and Yarranton G. A. (1973): Diversity richness and evenness during a primary sand dune succession at Grand Bend, Ontario.—Can. J. Bot. 51: 2401–2411.

Noble I. R. and Slatyer R.O. (1980): The use of vital attributes to predict successional changes in plant communities subject to recurrent disturbance.—Vegetatio 43: 5–21.

Numata M. (1982): Experimental studies on the early stages of secondary succession.—Vegetatio 48: 141–149.

O'Connor I. and Aarssen L. W. (1987): Species associations patterns in abandoned sand quarries.—Vegetatio 73: 101–109.

Oláh J. and Szegi J. (1979): Recultivation on the opencast piles of Gyöngyösvisonta.—Publ. of the Coalmines Firm of Mátraalja. Gyöngyös, 55 pp.

de Pablo C. L., Peco B., Galiano E. F., Nicolas J. P. and Pineda F. D. (1982): Space-time variability in mediterranean pastures analysed with diversity parameters.—Vegetatio 50: 113–125.

Palmer M. W. (1987): Variability in species richness within Minnesota oldfields: a use of the variance test.—Vegetatio 70: 61–64.

Peart D. R. (1989): Species interactions in a successional grassland. I. Seed rain and seedling recruitment.—J. Ecol. 77: 236–251.

Peart D. R. (1989): Species interactions in a successional grassland. II. Colonization of vegetated sites.—J. Ecol. 77: 252–266.

Peart D. R. (1989): Species interactions in a successional grassland. III. Effects of canopy gaps, gopher mounds and grazing on colonization.—J. Ecol. 77: 267–289.

Peet R. K. and Christensen N. L. (1980): Succession: a population process.—Vegetatio 43: 131–140.

Pickett S. T. A. (1982): Population patterns through twenty years of oldfield succession.—Vegetatio 49: 45–59.

Pickett S. T. A. and White P. S. [eds.](1985): The ecology of natural disturbance and patch dynamics.—Acad. Press, N. Y.

Podani J. (1984): Spatial processes in the analysis of vegetation. Theory and review.—Acta Bot. Acad. Sci. Hung. 30: 75–118.

Podani J. (1987): Computarized sampling in vegetation studies.—Coenoses 2: 9–18.

Prach K. (1984): Selected results of the study of succession on dumps from coal mining (Most Region, N.W. Bohemia).— Acta Bot. Slov. Acad. Sci. Slov., Ser. A., Suppl. 1., 257–261.

Prach K. (1986): Colonization of dumps from coal mining by higher plants.—Ecologia (ČSSR), Bratislava, 5: 421–424.

Prach K. (1986): Succession accross an environmental gradient.—Ecologia (ČSSR) 5: 425–430.

Prach K. (1987): Succession of vegetation on dumps from strip coal mining, N.W. Bohemia, Czechoslovakia.—Folia Geobot. Phytotax. 22: 339–354.

Rényi A. (1962): Wahrscheinlichkeitsrechnung, mit einem Anhang über Informationstheorie.— VEB Deutscher Verlag der Wissenschaften, Berlin.

Russell W. B. and LaRoi G. H. (1986): Natural vegetation and ecology of abandoned coal-mined land, Rocky Mountain Foothills, Alberta, Canada.—Can. J. Bot. 64: 1286–1298.

Szegi J., Oláh J., Fekete G., Halász T., Várallyay Gy. and Bartha S. (1988): Recultivation of the spoil banks created by open-cut mining activities in Hungary.—Ambio 17: 137–143.

Yarranton G. A. and Morrison R. G. (1974): Spatial dynamics of a primary succession: nucleation.—J. Ecol. 62: 417–428.

Watt A. S. (1947): Pattern and process in the plant community.—J. Ecol. 35: 1–22.

Wood D. M. and del Moral R. (1987): Mechanisms of early primary succession in subalpine habitats on Mount St. Helens.— Ecology 68: 780–790.

Krahulec F., Agnew A.D.Q., Agnew S. & Willems J.H. [eds.]: *Spatial processes in plant communities.* pp. 49–58.

The analysis of pattern change in a Hungarian sandy grassland

László Körmöczi and Andrea Balogh

Department of Botany, A.J. University of Szeged, H-6701 Szeged, P.O. Box 657, Hungary

Keywords: Community pattern, Ecotones, *Festuca pseudovina*, *Potentilla arenaria*, Transect

Abstract. Spatial and temporal dynamics of three grassland communities were investigated in sandy pasture of the Great Hungarian Plain. A 40 m long permanent transect was established to analyse the distribution of species, the overlap of their habitat and annual and seasonal alteration of distributions. The studied species showed characteristic localization along the transect, i.e. the elevation gradient. Continuous transition could be detected among species which were members of different associations. Only a few species had a marked border, and the edges of distribution of most species showed movement in the course of years. The unchanging border line were found at the steepest slope of the sand hill, or in the species which were situated in the deepest site of wind groove. We suggest that the causes of the changes of spatial distribution lie in the drought resistance of species.

Introduction

"The concept of vegetation as continuum with a changing species composition along environmental gradients arose in antithesis to the community-unit which stated that plant communities are natural units of coevolved species population forming homogeneous, discrete, and recognizable units." Austin (1985) summarizes thus the basic conflict existing between the two points of view of vegetation studies, and cites Whittaker's (1975) statement: "The broad overlap and scattered centres of species populations along a gradient imply that most communities intergrade continuously along environmental gradients, rather than forming distinct, clearly separated zones."

Studying the vegetation of an area of varied relief, the first problem for the researcher is how to distinguish the association patches in the patchy environment, because the communities intergrade not only in the abstract ecological space of gradients but also in topographical space. The transitional zone between adjacent associations also indicates environmental gradient(s). Another problem is how to follow the temporal movement of patch size and border, if the border line is ambiguous.

Sandy grassland associations in the Great Hungarian Plain gave opportunity to study these problems, since they form a mosaic and their existence in different zones of an elevation gradient is clearly recognizable.

49

The aim of our studies was to learn the behaviour of transitional zones of community patches and to follow their structural changes.

There emerged four questions to solve:

1. What is the distribution of the populations along the transect?
2. What is the overlap in the distribution of populations forming the adjacent communities?
3. How can the borders of associations be found and how sharp is the border line?
4. What kind of seasonal and annual changes can be recognized at the edge of communities and populations?

Materials and Methods

The investigation was carried out on a sandy grassland area in the Bugac region of Kiskunság National Park (Hungary; Fig. 1). Earlier this area was a pasture but it has been free of grazing since 1976. In previous researches we recognized four plant associations which form a mosaic-like grassland on a varied relief (Bodrogközy and Farkas 1981, Körmöczi et al. 1981).

We established a 40 m long permanent belt transect which runs throught two sand hills and between them a wind groove. The largest elevations difference is 2.5 m. The profile of the transect is given at Fig. 5.C. It was divided into 40 contiguous quadrats of 1 square meter. Three associations could be identified along the transect, *Festucetum vaginatae* at both ends, i.e. at the highest parts, *Potentillo-Festucetum pseudovinae* adjacent to the *Festucetum vaginatae* on the slopes, and *Molinio-Salicetum rosmarinifoliae* at the bottom of the "elevation gradient".

We called it the "elevation gradient" in the first approach, because this could be recognized immediately and the transect was laid down on this basis, but at the same time it also means a water regime gradient, a climatic gradient, gradients of several nutrients, etc.

In this paper only the structural properties of the vegetation are analysed in detail, but we also measured the background factors.

Records were made seasonally in 1985 and 1986, in spring and autumn 1987 and in spring 1988. Relevés were taken from each quadrat. Percentage cover values of species were recorded.

Data only from May 1987 were analysed in detail. The full data set of only two species are in this paper. For detecting abrupt changes in the composition of vegetation we represented the number of species entering and leaving the species set of successive quadrats. "Entering species" are those which are not found in the previous two quadrats and "leaving species" are not found in the next two quadrats (Molnár 1988).

Principal component analysis was used as an ordination technique to detect the multivariate relationships of plots. Only the first two axes are reported, since they are responsible for more then 50% of variance and they show clearly the gradient distribution (van der Maarel 1980).

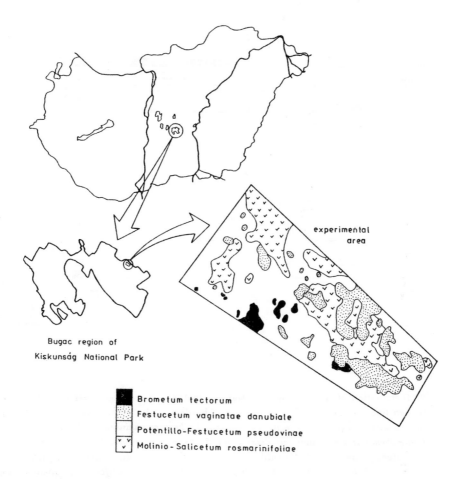

Bugac region of
Kiskunság National Park

■ Brometum tectorum
Festucetum vaginatae danubiale
Potentillo-Festucetum pseudovinae
Molinio-Salicetum rosmarinifoliae

Fig 1. The map of the experimental area.

Results and Discussion

Figs. 2 and 3 show the graphs of distribution of species in May 1985 and May 1987. The species are ordered on the basis of interspecific correlation (see Tables 1 and 2), but this order is well supported by cluster analyses. In both cases the species form three groups, within two of which they have positive interspecific correlations but in the third they have negative correlations or no correlations. The figures show the percentage cover classes of populations, and they give the structural basis for the visual estimation of the boundary zones as they are indicated.

The vegetation of the transect consisted of rather few species. In 1985 we found only 36 and in 1987 40. It can be seen from the figures that only 6–8 species were dominant and about half the species had wider distributions. Many species appeared sporadically or formed small patches.

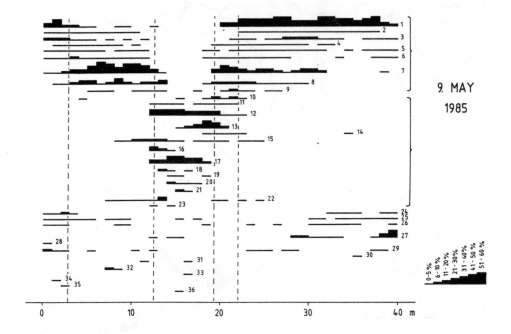

Fig. 2. Distribution of species along the transect in 1985. Percentage cover classes are shown. Broken lines indicate the first estimation of borders, and the brackets mean the groups of species with positive interspecific correlation. 1: *Festuca vaginata*; 2: *Medicago minima*; 3: *Poa bulbosa*; 4: *Cerastium semidecandrum*; 5: *Arenaria serpyllifolia*; 6: *Carex stenophylla*; 7: *Festuca pseudovina*; 8: *Potentilla arenaria*; 9: *Calamagrostis epigeios*; 10: *Achillea mille-folium*; 11: *Falcaria vulgaris*; 12: *Galium verum*; 13: *Poa angustifolia*; 14: *Silene otites*; 15: *Eryngium campestre*;16: *Crepis setosa*; 17: *Molinia caerulea*; 18: *Ononis spinosa*; 19: *Tarax-acum officinale*; 20: *Centaurea pannonica*; 21: *Salix rosmarinifolia*; 22: *Thymus degenianus*; 23: *Tragopogon dubius*; 24: *Secale silvestris*; 25: *Leontodon autumnalis*; 26: *Myosotis stricta*; 27: *Carex liparocarpos*; 28: *Cynodon dactylon*; 29: *Euphorbia segueriana*; 30: *Lithospermum arvense*; 31: *Onosma arenaria*; 32: *Poa pratensis*; 33: *Schoenus nigricans*; 34: *Senecio vernalis*; 35: *Syrenia cana*; 36: *Verbascum lychnitis*.

The species can be seen to show overlap and continuous transition among vegetation patches, mainly the species with positive correlation, and on the basis of these distributions we cannot see clearly the boundary of the communities.

Fig. 4. shows the distribution of six dominant species. By choosing only the dominants from the two sets, the situation of communities along the transect becomes clearer: we can more easily find the boundaries but not between the two dry grasslands. The range of dominant xeromorphic grass species overlapped widely, but the turn in dominance ratio

52

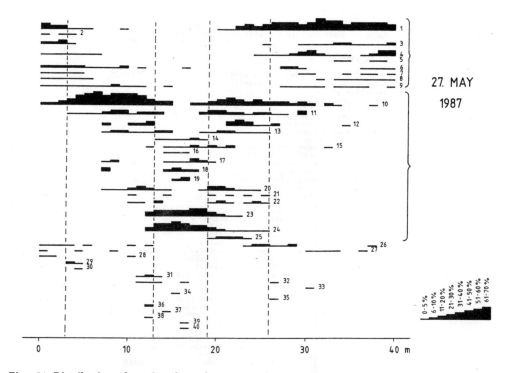

Fig. 3. Distribution of species along the transect in 1987. For the labellings see Fig. 2. *1: Festuca vaginata; 2: Arenaria serpyllifolia; 3: Secale silvestris; 4: Carex liparocarpos; 5: Bromus squarrosus; 6: Poa bulbosa; 7: Polygonum arenarium; 8: Kochia laniflora; 9: Cynodon dactylon; 10: Festuca pseudovina; 11: Potentilla arenaria; 12: Koeleria glauca; 13: Carex stenophylla; 14: Ononis spinosa; 15: Falcaria vulgaris; 16: Centaurea pannonica; 17: Holoschoenus vulgaris; 18: Poa angustifolia; 19: Salix rosmarinifolia; 20: Eryngium campestre; 21: Silene otites; 22: Thymus degenianus; 23: Galium verum; 24: Calamagrostis epigeios; 25: Achillea millefolium; 26: Euphorbia segueriana; 27: Myosotis stricta; 28: Equisetum ramosissimum; 29: Lithospermum arvense; 30: Stellaria media; 31: Dianthus pontederae; 32: Tragopogon dubius; 33: Veronica praecox; 34: Verbascum lychnitis; 35: Onosma arenaria; 36: Hieracium hoppeanum; 37: Taraxacum officinale; 38: Trinia ramosissima; 39: Polygala comosa; 40: Silene multiflora.*

may mark the border line.

Figure 5 shows the number of species entering (upper columns) and leaving (lower columns) the species set. As we can see, there are no very abrupt changes in the species composition, and the more marked alterations occur in the region of *Molinio-Salicetum rosmarinifoliae*. There is no clear distinction between the xeric grasslands because they have many species in common, and their environmental parameters (height, soil moisture, humus, calcium carbonate, soil hygroscopy) are similar. (The detailed description

53

Table 1. Interspecific correlations of species in 1985.

09.05.1985

```
:Achillea millefolium        % level of      positive  negative
 +:Falcaria vulgaris         significance
 xx:Galium verum
 +xx:Poa angustifolia            2.5             x         o
 xx+ :Silene otites              1               +         -
  ++  :Eryngium campestre        0.1             x         =
  +x   :Crepis setosa
  xx+ x+:Molinia caerulea
  +x   ++:Ononis spinosa
  +x     + :Taraxacum officinale
   ++   xxx:Centaurea pannonica
   x     +:Salix rosmarinifolia
     +      :Thymus degenianus
       x      :Thymus glabrescens
       x      :Tragopogon dubius
 o-- o =oo-  :Festuca vaginata
 === o = o   +:Medicago minima
 --- - =     +x:Poa bulbosa
        -      + :Cerastium semidecandrum
 o-   o=o -      x :Arenaria serpyllifolia
      - o      x+:Carex stenophylla
        -      +xx:Festuca pseudovina
               x +x:Potentilla arenaria
               x +:Calamagrostis epigeios
    o            :Secale silvestris
               -ob:Leontodon autumnalis
               - :Myosotis arvensis
          o -     :Carex liparocarpos
                  :Cynodon dactylon
                  :Euphorbia segueriana
                - :Lithospermum arvense
                  :Onosma arenaria
                  :Poa pratensis
                - - :Schoenus nigricans
                - - -:Senecio vernalis
                - - --:Syrenia cana
                - - ---:Verbascum lychnitis
```

and analysis of the effect of background factors are in preparation.) In comparison with another vegetation type (a marsh-meadow, where 49 species were found in spite of very favourable water conditions (Molnár 1988), in the sandy grassland the species exchange is more frequent, and the sum of entering and leaving species is larger, especially in the mesic community.

We tried to find the boundary of associations with multivariate analyses. PCA scatter diagrams of plots (Fig. 6) support again the difficulty of separating associations. Points which may be considered as members of *Molinio-Salicetum rosmarinifoliae* form a more or less distinct group, and we can mark the border (in 1985 between points 12–13 and 19–20, respectively), but we cannot distinguish between *Festucetum vaginatae* and *Potentillo-Festucetum pseudovinae*. In the second experimental year the pattern of the PCA scatter diagram is altered. The gradient distribution is recognizable, but the samples in mesic grassland are no longer well separated. This reflects successional processes, the shift of border zones and the alteration of vegetation structure that was caused by changing weather conditions (Körmöczi 1990).

The answers to the first three questions:

1. Most species show characteristic localisation along the transect in correlation with the elevation. These populations form continuous patches or some of them show

Table 2. Interspecific correlations of species in 1987.

```
27.05.1987

:Achillea millefolium          % level of      positive  negative
+:Calamagrostis epigeios        significance
xx:Galium verum
++ :Thymus degenianus              2.5             x         o
x  x:Silene otites                 1               +         -
++x+:Eryngium campestre            0.1             x         =
xxxx+x:Carex stenophylla
 +x      :Ononis spinosa
 xx      +:Falcaria vulgaris
  x      xx:Centaurea pannonica
  +      x++:Holoschoenus vulgaris
  x      xxxx:Poa angustifolia
  x x+      :Koeleria glauca
  + xx      x:Potentilla arenaria
     x x +  :Salix rosmarinifolia
       + :Festuca pseudovina
x==    -      :Cynodon dactylon
-o  -=o    -  x:Kochia laniflora
o   --        xx:Polygonum arenarium
=-     o      + x:Poa bulbosa
              x  :Bromus squarrosus
==   ==0- -   ++  :Carex liparocarpos
--   =-    =   xx  x:Secale silvestris
 -   oo=o =-   +x  x+:Festuca vaginata
               x    :Arenaria serpyllifolia
                   :Euphorbia segueriana
                   :Myosotis stricta
                   :Equisetum ramosissimum
                    :Lithospermum arvense
                    :Stellaria media
                   -:Taraxacum officinale
                   --:Verbascum lychnitis
                    :Veronica praecox
                   --- :Onosma arenaria
                     :Dianthus pontederae
                     x:Tragopogon dubius
                   --- -  :Hieracium hoppeanum
                   --- - +:Trinia ramosissima
                   --- - --:Polygala comosa
                   --- - --+:Silene multiflora
```

small discontinuities depending on the relief. The distributions developed according to the type of seed dispersal and vegetative spread shown by different species, and were modified by enviromnental parameters (primarily soil moisture) and competitive ability of other species.

2. The ranges of various species belonging to the same community were very different, and showed wide contacts and continuous transitions among communities by their overlaps. Even the dominant species of adjacent communities overlapped and formed transitional zones of different width between associations. This zone was 2–3 m wide between *Potentillo-Festucetum pseudovinae* and *Molinio-Salicetum rosmarinifolie*, but much wider between the two xeric grasslands. On this basis a marked border cannot be recognized.

3. To define the border line between two communities we have to use different methods, and the average of their results may give the best estimation of the position of borders.

These methods could be (1) visual estimation on the basis of vegetation structure, (2) distribution graphs of dominant species, (3) graphs of species exchanges and (4) multivariate analysis of species composition and environment parameters. The first two will not give sharp boundaries, but the others can, and the median of the given zone may be regarded as boundary between two communities.

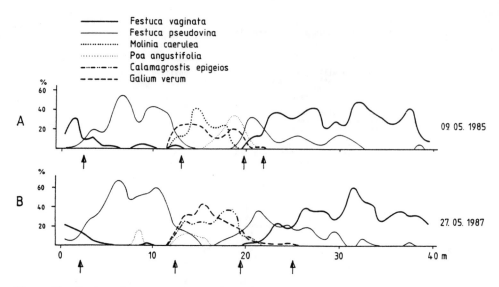

Fig. 4. Distribution of percentage cover values of six dominant species along the belt transect in 1985 (A) and 1987 (B). Arrows indicate the boundaries of associations.

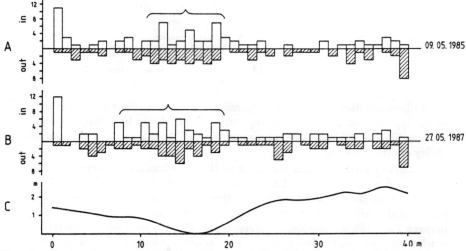

Fig. 5. Changes in the number of "entering" and "leaving" species in 1985 (A) and in 1987 (B) (open columns are entering and shaded columns are leaving species), and profile of the transect (C). Brackets indicate the area of *Molinio-Salicetum rosmarinifoliae*.

Why is it important to find this boundary? If somebody wants to draw a vegetation map, this procedure would be very time consuming and perhaps impossible.

But regarding the fourth question this method is useful when using one or several permanent transects to examine the dynamics of transitional zone of adjacent communities

56

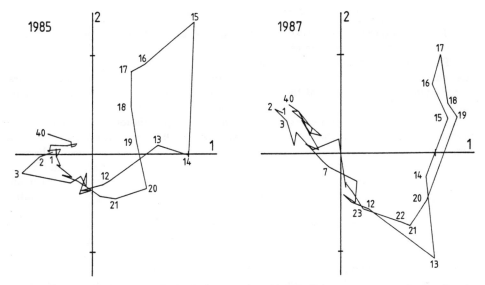

Fig. 6. PCA scatter diagrams of relevés from 1985 and 1987. Points are connected according to their position along the transect.

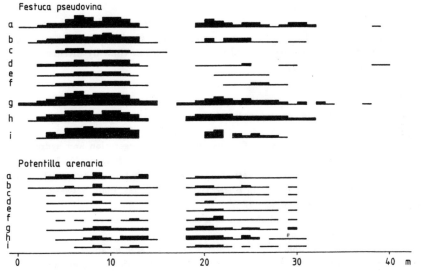

Fig. 7. Distribution of *Festuca pseudovina* and *Potentilla arenaria* along the transect in the experimental years (a: 09.05.1985; b: 27.08.1985; c: 05.10.1895; d: 11.05. 1986; e: 12.08.1986; f: 30.09.1986; g: 27.05.1987; h: 16.10.1987; i: 01.09.1988). Percentage cover classes are same as in Figs. 2. and 3.

and succesional changes.

The fourth question concerned the seasonal and annual changes of population distributions and of community boundary zone.

PCA scatter diagrams (Fig. 6) and associations in Tables 1 and 2 serve to demonstrate that changes in interrelationship of these vegetation samples have occurred between these sampling dates, we propose that these changes must be due to processes which act mainly throught the perennial species in our grasslands.

The causes of these processes are in the individual populations, and mainly in the perennials. We saw that the dominance of species changed between the two dates and Fig. 7 shows an example of distribution of two species at shorter time intervals. The cover dominance progression depended on the life cycles of the species and on background factors. The first species (*Festuca pseudovina*) shows one type of behaviour. Its appearance and disappearance fluctuated together with cover values, that is why the edge of its distribution is very diverse at both sides.

The other, *Potentilla arenaria*, however, shows a more or less unchanging range of distribution, though its cover values altered a little. An important feature of this second type can be seen at the left side of the graph, namely the gradual disappearance in the border zone of the two xeric grasslands. Maybe we can call the former species "mobile species" and the latter "slow species" in respect of response to changes in background factors.

Mobile species may possibly cause the vegetation fluctuations, and slow species the succesional changes, even if they are subdominant.

In this paper we investigated a phenomenon of vegetation, but did not study its relationship to backgroung factors. But it is evident that in these associations the short term changes are caused by environmental parameters rather than by intrinsic properties of populations (Körmöczi 1990).

References

Austin M.P. (1985): Continuum concept, ordination methods, and niche theory.—Ann.Rev. Ecol. Syst. 16:39–61.

Bodrogközy Gy. and Farkas Gy. (1981): Correlations between vegetation and hydroecology in the sandy grasslands of Kiskunság National Park.—Acta Biol. Szeged. 27:33-53.

Körmöczi L., Bodrogközy Gy. and Horváth I. (1981): Investigation of biological production and bioclimate of sandy grasslands in Bugac (Great Hungarian Plain between Danube and Tisza).—Acta Biol. Szeged. 27:55–69.

Körmöczi L. (1990): Drought-induced changes in a sandy grassland complex in the Great Hungarian Plain. (in preparation)

Molnár Zs. (1988): Investigation of plant communities of marsh-meadow Kis-tomalom with particular reference to contacts of associations.—Thesis (Szeged)[in Hungarian].

van der Maarel E. (1980): On the interpretability of ordination diagrams.—Vegetatio 42:43-45.

Whittaker R.H. (1975): Communities and ecosystems.—2nd ed. McMillan, New York.

Krahulec F., Agnew A.D.Q., Agnew S., Willems J.H. [eds.]: *Spatial processes in plant communities.* pp. 59–69.

Spatial changes in herb layer in a beech forest/clear-cut area ecotone from northern Hungary

Ilona Mészáros

Botanical Department of Lajos Kossuth University, H-4010 Debrecen, Hungary

Keywords: Ecotone, Diversity, Belt transect, Clear-cutting, *Melico-Fagetum*.

Abstract. This paper describes the trends in spatial and temporal changes in the herbaceous vegetation of a beech forest/clear-cut area ecotone during the first years after the deforestation. In the field studies the "belt transect method" was applied (1 × 32 m transect perpendicular to the forest edge). Shortly after the clear cutting a significant structural rearrangement of the herb layer was observed in the forest edge. Although the species composition and the number of species changed slightly, a considerable increase in the density of aboveground shoots took place simultaneously which was mostly attributed to the multiplication of the shoots of some forest species. The strong alteration of the dominance pattern resulted in a great decrease of the species-shoot number diversity in comparison with the forest interior. From the results of cluster analysis it could be concluded that the so called "edge effect" occurred in a 8–10 m wide transition zone between the beech forest stand and the deforested area during the study period.

Introduction

Nowadays in landscape management there is an increasing need for understanding the processes taking place in ecotones. Ecotone is traditionally defined as a transition zone between plant communities (Clements 1928, Odum 1971, Weaver and Clements 1929). It may vary in nature from abrupt discontinuities to broad and gentle gradients depending on the scale of observation (Krebs 1972, van der Maarel 1966, 1976, Jagomägi *et al.* 1988). In a landscape its most striking and easily recognizable forms are the forest edges regarded as small-scale forest ecotones which occur as contact zones between forest stands and treeless areas and are created and/or maintained by human activity or natural processes. Although the forest edges usually extend to a long and narrow zone, they act as a biofilter or "semi-permeable membrane" in the flow of energy, materials and organisms (Hansen *et al.* 1988, Ranney 1977, Zólyomi 1987). Several authors (Geiger 1950, Jakucs 1972, Ranney 1977, Wales 1972) emphasize that forest edges may considerably buffer the forest interior against disturbances coming from the open area and they make much account of their vertical and horizontal structures in this function. Since only sporadic findings are available concerning this aspect, in landscape ecology more research is needed on these edge-areas as they may have important implications for the management of disturbance.

59

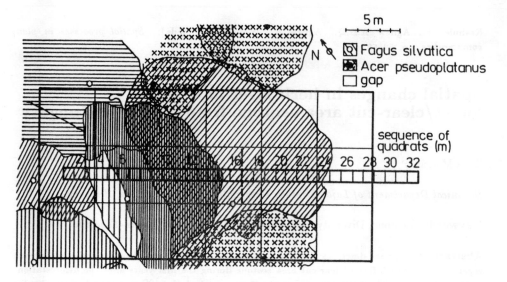

Fig. 1. Location of transect and sampling quadrats and foliage cover of trees above them.

The principal aim of this study is to investigate the spatial changes in the herb layer across a beech forest/clear-cut area ecotone 5 years after the deforestation. Specifically, the objective is to evaluate the changes in species composition, shoot density, diversity and dominance pattern of the herb layer in the contact zone and determine the character and the width of the transition zone. Since ecotones have special significance both theoretically and practically, our intention is to provide further material to understand better the above mentioned problems.

Study area

The studies were performed in July 1985 on the research area of "Rejtek Project", coordinated by the Ecological Department of Lajos Kossuth University, Debrecen, and described by Jakucs (1987). The base rock of the area is limestone which is covered by shallow black rendzina soil.

Till the winter of 1980/81 the largest part of the area was covered by two subassociations of old sub-montane beech forest (*Melico-Fagetum melicetosum uniflorae* and *Melico-Fagetum asperuletosum odoratae*) when 4.5 ha of the forest was clear-cut. The herbaceous vegetation has been studied in the contact zone of the remaining fragment of *Melico-Fagetum melicetosum uniflorae* subassociation and the deforested area on the central "plateau" of the research area (Jakucs 1987).

In the tree layer of the forest the average canopy cover is 85%, the dominant species is *Fagus sylvatica* which is sporadically accompanied by *Acer pseudoplatanus*, *Acer platanoides* and *Carpinus betulus* (Jakucs 1987, and Fig. 1 here). The shrub layer is practically absent, only the samplings of tree species can be found. In the herb layer, which is rather poor, the characteristic species of *Fagetalia* prevail.

60

Methods

Field investigations

The vegetation was investigated in a permanent belt transect (1 m wide and 32 m long) placed perpendicular to the forest edge. The transect was composed of contiguous 1 m² quadrats which were numbered in running order beginning in the forest stand and ending in the clear-cut area. Every second quadrat was sampled. The quadrats were again divided into small quadrats of 25 × 25 cm by means of a wooden frame. Within these small sampling plots all herb layer species were recorded, and their aboveground shoots were counted. Then the data from these small quadrats are summed to express the results per 1 m².

Data analysis

To describe the rate of change in vegetation differential profiles were applied (Hobbs 1986) which were derived for the transect using the coefficient of difference (CD) calculated by Czekanowski-index (Czekanovski 1913) on the basis of the presence and absence of species and the aboveground shoot density of each species in the sampling quadrats, respectively. The values of CD were obtained by solving the equation successively for each quadrats along the transect in comparison with the end quadrat extending to the clear cut area. Then the values were subjected to three-point moving averaging (Kovacsics 1979) to reduce the amount of point to point variation. Since this method does not inform about the width of the transition zone, cluster analysis was also applied (Orlóci 1978) on the basis of shoot number of each species in the quadrats.

To measure the diversity several indices are available. In the permanent study the species-shoot number diversity was calculated by Shannon-formula (H_S; Shannon and Weaver 1949), Brillouin-index (H_B; Brillouin 1962) and Hill's diversity indices ($N_{1/2}$, N_1; Hill 1973). The equitability was estimated on the basis of the equation suggested by Pielou (1975).

Results

The number of species in the sampling quadrats changed between 14 and 33 (Fig. 2). In the first quadrats of the transect moderately low values of number of species were found (17–23) and mostly typical beech forest species were present (e.g. *Asperula odorata, Dentaria bulbifera, Melica uniflora, Mercurialis perennis, Stellaria holostea*). The quadrats 12–20 representing the forest edge were generally floristically poorer, the number of species fluctuated between 14 and 20. Despite the relatively low number of species a considerable turnover of species occurred on this segment of the transect. Beside the loss of certain forest species several new colonizing species (e.g. *Cirsium arvense, Clematis vitalba, Geranium robertianum, Geum urbanum, Polygonum dumetorum, Urtica dioica*) appeared. In quadrats 22–32 belonging to the clear-cut area the number of species reached high values (26–33) and the proportion of new colonizing species was higher than in the forest edge. Several additional species were recorded in these quadrats (e.g. *Cirsium vulgare, Eupatorium cannabinum, Rosa* sp., *Rubus idaeus*).

The total density of aboveground shoots of herb layer species showed a large fluctuations along the transect (Figure 3). It was very low in the internal ("forest") quadrats and

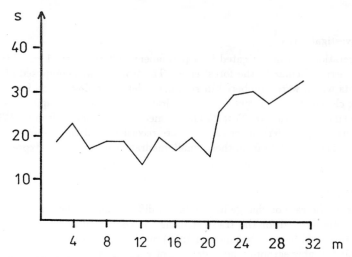

Fig. 2. Number of herb layer species in sampling quadrats of the transect.

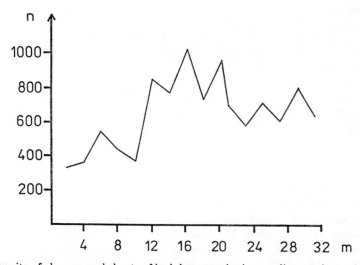

Fig. 3. Density of aboveground shoots of herb layer species in sampling quadrats of the transect.

changed between 330 and 560. In the forest edge, where the species composition and the number of species changed slightly less than on the deforested area, a significant increase in the shoot density took place simultaneously. The density was twice as high as in the forest. It was mainly due to the multiplication of the aboveground sprouts of some forest species, especially of *Melica uniflora*, and only in a small degree to the colonization of new invading species.

Lower values of shoot density (615–838) occurred again in the quadrats belonging to

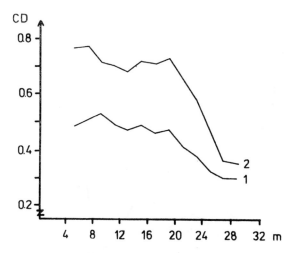

Fig. 4. Differential profiles constructed on the basis of presence/absence (1) and shoot density (2) of each species in the sampling quadrats.

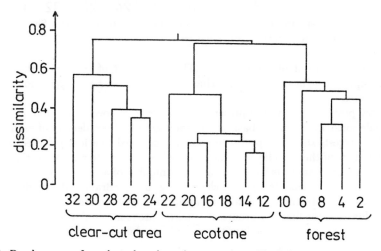

Fig. 5. Dendrogram of quadrats based on shoot number of herb layer species.

the clear-cut area (Fig. 3) which can be explained by the loss of importance of *Melica uniflora* and relatively large individuals of some colonizing species.

The differential profile method and the cluster analysis used here for evaluating the spatial changes in the structural characteristics of the herb layer along the transect gave somewhat different results (Figs. 4 and 5).

Differential profiles (Fig. 4) begin with high values of dissimilarity and have very small steepness in their first section indicating slight and gradual alteration in the internal (forest) quadrats. From the quadrat 22 a strong decline of dissimilarity appears which

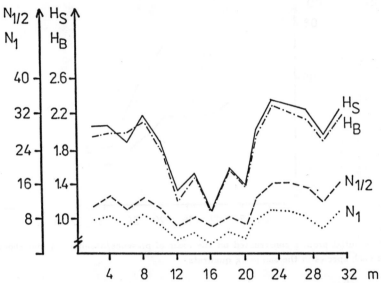

Fig. 6. Spatial changes in species shoot number diversity along the transect (H_S: Shannon-index; H_B: Brillouin index; $N_{1/2}$ and N_1: Hill's indices).

goes on to the end quadrat of the transect. This steep part of the differential profiles reflects an abrupt change in the structural characteristics of the herb layer coming nearer the deforested area.

Differential profile based on the presence and absence of species is running "higher" than the profile constructed on the basis of the aboveground shoot density of each species (Fig. 4). This suggests that the species composition changed to a lesser degree than the abundance relations of the herb layer in the forest edge 5 years after deforestation.

When using cluster analysis based on the shoot number of all species in the sampling quadrats three transect segments separate clearly (Fig. 5): quadrats 2–10, quadrats 12–22 and quadrats 24–32 representing the forest interior, the ecotone and the clear-cut area, respectively. However, the first two groups of quadrats are more similar to each other than to the third group and are arranged into the same large cluster.

Diversity and dominance/diversity curves

The different indices applied show similar tendencies of change in the species-shoot number diversity of the herb layer along the transect (Fig. 6). The curves of all the four indices are U-shaped curves due to the very low values of diversity in the quadrats 12–20 belonging to the ecotone. H_S and H_B indices, however, indicate larger fluctuation of diversity than the Hill's indices due to their different sensitivity to the large, moderate and small frequencies as shown by del Moral and Fleming (1980), McIntosh (1967), Nosek (1976), and Pielou (1975). While the values of four indices differ considerably in the forest and in the clear-cut area their small discrepancies can be observed in the ecotone. The great decrease of species shoot number diversity in the forest edge in comparison with

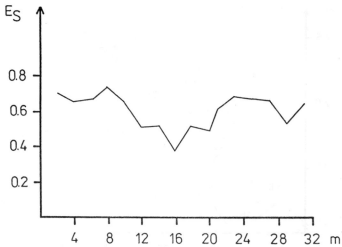

Fig. 7. Spatial changes in species shoot number equitability along the transect.

the adjoining areas resulted from the significant increase in the dominance and the very uneven distribution of aboveground shoots of species as supported by the low values of equitability (Fig. 7).

The diversity showed strong positive correlation with the number of species and strong negative correlation with the density of shoots along the transsect (Fig. 8). The figures reflects the separation of clouds of points representing three segments of the transect (the forest, the ecotone and the clear-cut area) which agree well with the grouping of quadrats obtained by cluster analysis (Fig. 5).

To examine the possible causes of differences in the diversity and dominance relations among the species in the three segments of transect dominance/diversity curves were constructed. The relative shoot densities of herb layer species occurring in the individual segments were used as importance values of species (Fig. 9).

The shape of the curve for the forest is approximately linear, the steepness of the upper and lower parts does not differ significantly. This type of dominance/diversity curve often describes the understory layers of forest communities (Whittaker and Woodwell 1968) and is generally attributed to the weak interspecific competition and the direct contact of niches of species (Whittaker 1965).

In the ecotone a considerable change in the shape of dominance/diversity curve can be observed (Fig. 9). The uppermost section of the curve becomes longer and very steep indicating high dominance of a single species, namely *Melica uniflora*. It is worth mentioning that in the forest edge the importance value of this species increased 10 times as compared to the forest. Other species such as *Viola hirta, Brachypodium silvaticum* are also dominant ones in the ecotone. The lower section of the curve for ecotone is less steep with even distribution of species of medium and low importance values. It seems likely that the shape of this curve is a result of a strong competition and a high control of resources available by one and/or some dominant species (Naveh and Whittaker 1979,

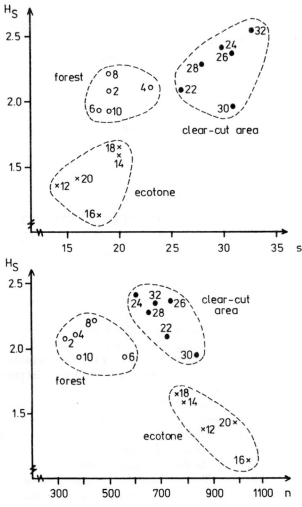

Fig. 8. Relationship between diversity H_S and: a — number of species (s) r = 0.746, $P < 0.1\%$; b — density of aboveground shoots (n) r = 0.620, $P < 5\%$.

Précsényi 1981).

The curve for the clear-cut area becomes more flattened (Fig. 9). The uppermost section of the curve is not so steep as in case of the ecotone and begins lower importance value. An increase in the number of dominant species can be observed. In its second section with lower steepness there are many subordinate species, first the species of medium importance are distributed approximately evenly and then the species of low importance values show coarser distribution. The shape of the curve refers to strong competition in

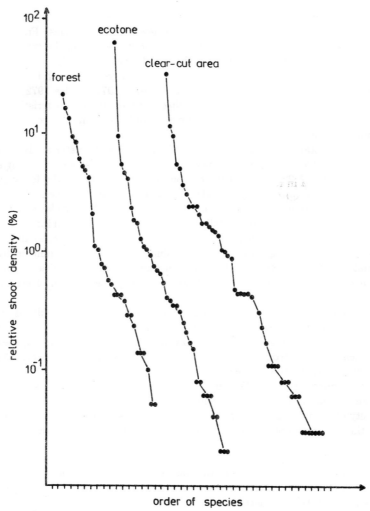

Fig. 9. Dominance/diversity curves for the three different segments of the transect.

the clear-cut area.

Discussion

Ecotones between forest stands and treeless areas generally offer favourable conditions for both mesofrequent, shade-tolerant species and xerofrequent, shade-intolerant species through their special microclimate (Geiger 1950, Mészáros *et al.* 1981, Ranney 1977, Wales 1972). The present study has demonstrated that a slight change occurred in the number of herb layer species in the ecotone between the beech forest and the clear-cut area 5 years after deforestation but it was accompanied by the considerable alteration of the shoot density (Figs. 2 and 3). The edge of the forest was a good habitat for both the survival of many forest species and the colonization of new invading species. From the variation of the density of aboveground shoots of herb layer species along the transect (Fig. 3) it can be concluded that the conditions of the forest ecotone greatly favour the vegetative reproduction and/or the aggregation of species which agree well with the observations on different kinds of forest edges (Jakucs 1972, Ranney 1977, Wales 1972). In our study it was especially conspicuous in case of *Melica uniflora*, the characteristic species of the subassociation, which multiplicated its aboveground shoots in the ecotone and became predominant species of the herb layer.

The resuls obtained by differentiall profile method and cluster analysis (Figs. 4 and 5) give evidence that (i) the ecotone between the beech forest and clear-cut area is 8–10 m wide (quadrats 12–22), and (ii) the ecotone is more similar to the forest interior than to the clear-cut area in respect of the structural characteristic of the herb layer.

It seems likely that the changed dominance pattern of the herb layer in the ecotone in comparison with the forest is principally attributed to the successful interspecific competition of *Melica uniflora*. We conclude that the strong dominance (Fig. 9) and the relatively low number of species (Fig. 2) result in low values of species-shoot number diversity (Fig. 6) and equitability (Fig. 8) of the herb layer in the ecotone. In addition trends of spatial change in diversity and equitability indicate very sharp separation of ecotone from the forest interior and the clear-cut area.

In summary, several aspects of "edge effect" (Ghiselin 1977) could be recognised in the ecotone between the beech forest and the clear-cut area already 5 years after the deforestation such as the successfull colonization of new invading species, the high values of shoot density, the strong dominance resulted in low values of species-shoot number diversity and generally the coarse spatial change in structural characteristics of herb layer between the adjoining areas. We can conclude that it is essential to pay attention to the maintenance of newly created edges which can play an important role in influencing the biodiversity of a landscape mainly through their effects on the spreading of plant species between the adjoining areas and act as a barier for many new colonizing species.

References

Brillouin L. (1962): Science and information theory. 2nd ed.—Academic Press, New York.

Clements F.E. (1928): Plant succession and indication. 2nd ed.—Washington.

Czekanowski J. (1913): Zarys metod statystycznych.—Warsaw.

del Moral R. and Fleming R.S. (1980): Structure of coniferous forest communities in Western Washington: diversity and ecotope properties.—Vegetatio 41: 143–153.

Geiger R. (1950): Climate near the ground.—Harvard Univ. Press, Cambridge.

Ghiselin J. (1977): Analyzing ecotones to predict biotic productivity.—Environmental Management 1: 235–238.

Hansen A.J., di Castri F. and Naiman R.J. (1988): Ecotones: what and why?—Biology International, Special Issue 17: 9–47.

Hill M.O. (1973): Diversity and evenness: a unifying notation and its consequences.—Ecology 54:427–432.

Hobbs E.R. (1986): Characterizing the boundary between California annual grassland and coastal sage scrub with differential profiles.—Vegetatio 65: 115–126.

Jagomägi J., Külvik M., Mander Ü. and Jacuchno V. (1988): The structural-functional role of ecotones in the landscape.—Ecológia (ČSSR) 7: 81–94.

Jakucs P. (1972): Dynamische Verbindung der Wälder und Rasen. Quantitative und qualitative Untersuchungen über die synökologischen, phytozönologischen und strukturellen Verhältnisse der Waldsäume.—Akadémiai Kiadó, Budapest.

Jakucs P. (1987): "Rejtek Project". Ecological researches for favourable regeneration of our beech forests covering shallow soil derived from limestone.—Acta Biol. Debr. 19: 5–13.

Kovacsics J. (1979): Statisztika.. [Statistics].—Tankönyvkiadó, Budapest.

Krebs C.J. (1972): Ecology. The experimental analysis of distribution and abundance.—New York.

McIntosh R.P. (1967): The continuum concept of vegetation.—Bot. Rev. 33: 130–187.

Mészáros I., Jakucs P. and Précsényi I. (1981): Diversity and niche changes of shrub species within forest margin.—Acta Bot. Acad. Sci. Hung. 27: 421–437.

Naveh Z. and Whittaker R.H. (1979): Structural and floristic diversity of shrublands and woodlands in Northern Israel and other mediterranean areas.—Vegetatio 41: 171–190.

Nosek J.N. (1976): Comparative analysis of some diversity functions under different conditions of sampling in sandy meadow.—Acta Bot. Acad. Sci. Hung. 22: 415–436.

Odum E.P. (1971): Fundamentals of ecology.—Saunders, Philadelphia, London, Toronto.

Orlóci L. (1978): Multivariate analysis in vegetation research.—Junk, The Hague.

Pielou E.C. (1975): Ecological diversity.—Wiley and Sons, New York.

Précsényi I. (1981): Changes in the diversity of the vegetation during succession.—Acta Bot. Acad. Sci. Hung. 27: 189–198.

Ranney J.W. (1977): Forest island edges—their structure, development and importance to regional forest ecosystem dynamics.—Environ. Sci. Div., Publ. No. 1069, Oak Ridge Nat. Lab., Oak Ridge.

Shannon C.E. and Weaver W. (1949): The mathematical theory of communication.—University of Illinois Press, Urbana.

van der Maarel E. (1966): Dutch studies on coastal sand dune vegetation, especially the Delta region.—Wentia 15: 47–82.

van der Maarel E. (1976): On the establishment of plant community boundaries.—Ber. Deutsch. Bot. Ges. 89: 415–443.

Wales B.A. (1972): Vegetation analysis of north and south edges in nature oak-hickory forest.—Ecol. Monographs 42: 451–471.

Weaver J.E. and Clements F.E. (1929): Plant ecology.—McGraw-Hill Book Comp., Inc., New York.

Whittaker R.H. (1965): Dominance and diversity in land plant communities.—Science 147: 250–260.

Whittaker R.H. and Woodwell G.M. (1968): Structure, production and diversity of the oak-pine at Brookhaven, New York. Ecology 57: 154–175.

Zólyomi B. (1987): Coenotone, ecotone and their role in preserving relic species.—Acta Bot. Hung. 33: 3–18.

Krahulec F., Agnew A.D.Q., Agnew S., Willems J.H. [eds.]: *Spatial processes in plant communities.* pp. 71–82.

Comparison of transect methods for the analysis of spatial pattern

Jan Lepš

Department of Biomathematics, Biological Research Centre, Branišovská 31, CS-370 05 České Budějovice, Czechoslovakia.

Keywords: Pattern analysis, Local quadrat variance, Transect methods, Euclidean distance, Community pattern.

Abstract. Five methods for the analysis of spatial pattern based on transect data were compared: the classical blocked quadrat method, the stepped blocked quadrat method, the two term local quadrat variance method, the three term local quadrat variance method and paired quadrat variance method. Both artificial and field data were used. The three term local quadrat variance performs best. The possibility of using this method for study of spatial relationships of two populations was shown. A new method of the analysis of pattern of the whole community based on squared Euclidean distance of neighbouring blocks was proposed.

Introduction

The spatial pattern of a population or of a community is considered to be an important structural characteristic (Greig-Smith 1979, Kershaw and Looney 1986, Pielou 1974). The spatial pattern of a population may be analyzed in several ways (distance methods, counts of individuals in randomly placed quadrats, in continuous quadrats, in transects). Examples are in many textbooks of quantitative or mathematical ecology. There is some disagreement about the efficiency of particular methods. The methods with good theoretical background are often difficult to apply in the field. For example the statistically well founded method proposed by Ripley (1987) is based on the distances from each individual to all other individuals in the population. Obviously, such methods are useful for analysis of digitized vegetation maps (if all individuals are mapped), but hardly applicable in the field. In analysis of species with vegetative propagation, the definition of an individual may be difficult. The field sampling for all these methods is tedious and laborious. The transect methods are based on counts, frequency or cover estimates in sampling units arranged in a belt transect. They are not based on particular statistical models - in fact, their only aspiration is to find scales of aggregations in a population. If more populations are sampled by a transect, their spatial relationships of them may be analyzed. The analysis may be broadened to several species or to a whole community. Various methods have been proposed for the analysis of transect data (Kershaw 1961, Usher 1969, Hill 1973, Ludwig

71

and Goodall 1978). The aim of this paper is to apply these methods to artificial data and to real data from old field communities and to compare their performance.

Material and Methods

Two data types were used – an artificially constructed transect and field data from old fields in Bohemian Karst. The artificial transect was constructed to simulate population with two scales of aggregation (at dimensions 7 and 40) and with a trend along the transect. The field data were obtained in the course of broader successional study in the Bohemian Karst, Central Bohemia (Osbornová *et al.* 1990). The data from the 7 yrs old field are used in this paper. The transect was performed by a long tape. Frequency of the five most common species (*Agropyron repens, Artemisia vulgaris, Sinapis arvensis, Convolvulus arvensis, Galium aparine*) was determined in four 2 cm subunits within a basic 8 cm unit. Three transects, each of the length of 128 basic units (i.e. 1024 cm) were analyzed. A species was considered to be present in the subunit if any of its aboveground parts crossed the tape. The sampling was performed in 1976.

In this paper, the notation and names of methods are according to Hill (1973), with some minor changes. I use the term *mean square* instead of *variance* to distinguish from the variance in ordinary statistical meaning. It is supposed that the data form a single transect consisting of $n + m$ basic units, where n is integer power of 2. The transect is represented by a sequence of species quantities in particular sampling units x_i:

$$x_1, x_2, ..., x_n, x_{n+1}, ..., x_{n+m}.$$

The following methods were compared:

1. *Blocked-Quadrat Variance Method.* (Bqv). Classical method proposed in fifties (Greig-Smith 1952, Kershaw 1957). The length of the transect is supposed to be an integer power of 2. The mean square at block-size 1 is

$$MS(1) = \frac{1}{n} \left((x_1 - x_2)^2 + (x_3 - x_4)^2 + ... + (x_{n-1} - x_n)^2 \right) \quad (1)$$

It is an average of $n/2$ non-overlapping terms $\frac{(x_i - x_{i+1})^2}{2}$. At block size two, mean square is an average of $n/4$ non-overlapping terms

$$\frac{(x_i + x_{i+1} - x_{i+2} - x_{i+3})^2}{4}.$$

Similarly, the MS is defined for any block-size which is an integer power of 2.

2. *Stepped Blocked-Quadrat Variance Method.* (Sbqv). This method (proposed by Usher 1969) is an extension of the previous one. m quadrats are added to the original transect. The starting point is step by step 'shifted' to first $m + 1$ positions on the transect and the result is an average of $m + 1$ Bqv analyses. At block-size 1, the mean square is defined

$$MS(1) = \frac{1}{m+1} \sum_{i=0}^{m} \frac{1}{n} \left((x_{1+i} - x_{2+i})^2 + (x_{3+i} - x_{4+i})^2 + ... + (x_{n-1+i} - x_{n+i})^2 \right)$$

$$(2)$$

and analogously at higher block sizes. Defined for block sizes which are an integer power of two.

3. *Two Term Local Quadrat Variance Method.* (Ttlqv). The method was proposed by Hill (1973). The mean square is defined as an average of overlapping terms, for example, for block size 1:

$$\text{MS}(1) = \frac{1}{m+n-1}\left(\frac{(x_1 - x_2)^2}{2} + \frac{(x_2 - x_3)^2}{2} + ... + \frac{(x_{n+m-1} - x_{n+m})^2}{2}\right) \quad (3)$$

There are no restrictions on the transect length. The mean square is defined analogously for any block size; for example, at block-size 3, the MS is defined as the average of overlapping terms

$$\frac{(x_1 + x_2 + x_3 - x_4 - x_5 - x_6)^2}{6}, \frac{(x_2 + x_3 + x_4 - x_5 - x_6 - x_7)^2}{6}, ...\text{etc.}$$

4. *Three Term Local Quadrat Variance* (T3lqv). This method proposed by Hill (1973) aimed to 'detrend' the analysis, that is to avoid the possible influence of trend (steady decrease or increase of abundance of a species along the transect) on the results. At block size 1, the MS is defined

$$\text{MS}(1) = \frac{1}{m+n-2}\left(\frac{(x_1 - 2x_2 + x_3)^2}{6} + \frac{(x_2 - 2x_3 + x_4)^2}{6} + ... \right. $$
$$\left. ... + \frac{(x_{n+m-2} - 2x_{n+m-1} + x_{n+m})^2}{6}\right) \quad (4)$$

and analogously for higher block sizes.

5. *Paired Quadrat Variance Method* (Pqv). Goodall (1974) has suggested a method of estimating spatial pattern on the basis of differences of species abundance in randomly chosen pairs of quadrats of given spacing. No blocks were used in this method. Following Ludwig (1979), we used all possible pairs of given spacing (instead of a randomly chosen subset). The mean square at spacing k is defined as an average of all possible terms $\frac{(x_i - x_{i+k})^2}{2}$, it is

$$\text{MS}(k) = \frac{1}{n+m-k}\sum_{i=1}^{n+m-k}\frac{(x_i - x_{i+k})^2}{2} \quad (5).$$

The statistical testing is difficult. The only tests proposed are for the Bqv method (Mead 1974) and for paired-quadrat methods. We consider these method as purely descriptive ones. Nevertheless, it is important to take into account that the reliability decreases with the decreasing number of averaged terms in each method. (Note that in Ttlqv and T3lqv, the particular terms are overlapping and so not independent variables). Hence, if the block size is close to half the length of the transect, the value of MS is the squared difference of two independent terms — so it has nearly no reliability. Consequently, the

73

analysis was always stopped at a block size considerably lower than a half of the transect length. If more transects of the same length are analyzed in a community and there is good reason to expect the same pattern in the whole area, analyses from particular transects may be averaged. We have applied this for field data.

If more species are analyzed by a transect, their spatial relationship may be investigated. The method proposed by Kershaw (1961) is based on the fact that the MS of two independently distributed species is

$$MS_{x+y} = MS_x + MS_y, \tag{6}$$

at any block size (similarly as with ordinary variances). If the two species are positively associated, the MS of their total is higher, if negatively associated, the MS is lower. Based on this, COMS may be defined as a measure of association analogous to ordinary covariance

$$COMS_{x,y} = \frac{MS_{x+y} - MS_x - MS_y}{2} \tag{7}$$

at any block size and an analog of correlation coefficient

$$rMS_{x,y} = \frac{COMS_{x,y}}{\sqrt{MS_x MS_y}}. \tag{8}$$

The pattern of the whole community may be investigated in several ways. For example, Galiano (1983) and Gibson and Greig-Smith (1986) used first an ordination to obtain an integrated score for each quadrat and the scores were then analyzed by variance methods. It seems that in the ordination some information might be lost. In most methods, the squared differences of species abundance are used in particular terms. If squared Euclidean distances are used instead, the community pattern may be analyzed directly. For example, for the Ttlqv, the mean square at block size one is defined

$$MS(1) = \frac{1}{m+n-1} \left(\frac{ED_{1,2}^2}{2} + \frac{ED_{2,3}^2}{2} + ... + \frac{ED_{m+n-1,m+n}^2}{2} \right), \tag{9}$$

where $ED_{i,j}$ is the Euclidean distance between samples i and j, as commonly used in numerical classifications (Orlóci 1978). It may be easily shown that the above defined pattern of the whole community is the total of patterns of its constituent species. In our study, only dominant species were recorded and consequently, the spatial pattern of a community is determined only on this basis.

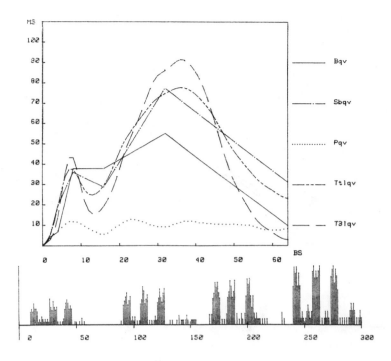

Fig. 1. Results of pattern analysis of artificial data (single population analyzed by five various methods). The transect is displayed below.

Results

The analysis of artificial data. — The results of pattern analysis of artificial data are displayed in Fig. 1. The T3lqv method performs best. It exactly determined both scales of aggregation and is not influenced by the trend. The Ttlqv method also performs well, but the peaks are not so sharp. The Sbqv method determined the scales within the accuracy given by integer powers of 2. The original Bqv method does not show a sharp peak at the lower scale. The Pqv method performs poorly, in particular, it was not able to detect aggregation on the higher scale.

The analysis of spatial relationships of two artificial populations is shown in Fig. 2. The populations are constructed to have similar pattern with two scales of aggregation (at BS 7 and 40); at the smaller scale, the populations are negatively associated, at the larger scale, positively associated. The T3lqv method was used and the relationships are recognized quite well.

The analysis of real populations. — Analyses of five species from the old field are displayed in Figs 3,4 and 5. As the length of the transect is an integer power of 2, (i.e. $m = 0$), the Sbqv analysis is reduced to the Bqv. The Pqv analysis seems to provide no interpretable description of the species pattern, as in the artificial case. The patterns of particular species are not as obvious as in the artificial example. As a consequence, the differences

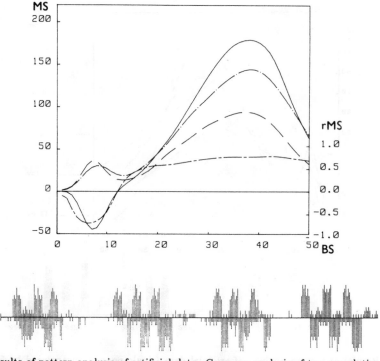

Fig. 2. Results of pattern analysis of artificial data. Common analysis of two populations by the T3lqv method — · — · — · species 1, — — — species 2, — COMS, — · — · — · rMS. The transects are displayed below.

between particular methods are greater. Particularly, the Bqv analysis at the largest block size often differs from both Ttlqv and T3lqv analyses. It is probably caused by low number of terms that are averaged. Note that The Bqv is *not* consistently higher or consistently lower. The results show that the limitation of the Bqv method to integer powers of 2 is a serious drawback. As in the artificial example, the T3lqv method seems to perform best. Important information may be obtained by comparison of Ttlqv and T3lqv. Their difference corresponds to the influence of trend (or of aggregation on higher scales).

Agropyron repens (Fig. 3) formed relatively homogeneous growth. The peak at the size 11 units (88 cm) partially corresponds to gaps in the continuous bed (see Pielou 1969, p. 106, who showed that the peak corresponds to gaps as well as to patches). At higher block sizes (about 30) there is some variation in the density of growth.

Artemisia vulgaris (Fig. 3) is very patchily distributed. The peak at block size 6 (48 cm) is not very distinct and is recognized only by the T3lqv method. It corresponds to basic 'morphological' patches of the species. There is a pronounced peak at BS of about 20 units (and corresponding at 16 units for Bqv); increase in Ttlqv suggests heterogeneity at higher scales as well. Their origin is not clear — probably environmental heterogeneity (microtopography) and spatial dynamics of the species.

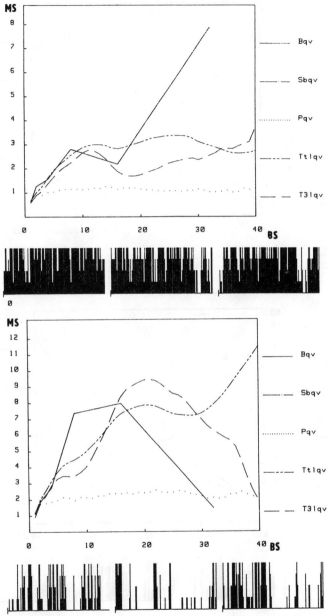

Fig. 3. Pattern analysis of real populations. Above — *Agropyron repens*, below — *Artemisia vulgaris*. Transects are displayed below.

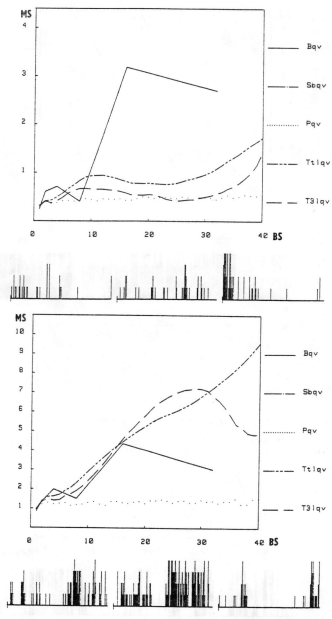

Fig. 4. Pattern analysis of real populations. Above — *Sinapis arvensis*, below — *Galium aparine*. Transects are displayed below.

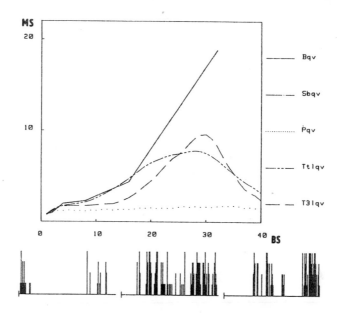

Fig. 5. Pattern analysis of *Convolvulus arvensis*. Transects are displayed below.

Sinapis arvensis (Fig. 4) does not show any distinct peak. The size of aggregations of this species is determined by the size of gaps in growth of competitively stronger perennials, which have variable size and are caused by minor disturbances and microtopography.

Galium aparine (Fig. 4) and *Convolvulus arvensis* (Fig. 5) are both strikingly clumped. Both of them show aggregations at block size of about 30 (240 cm).

The pattern of the whole community (Fig. 6) shows a peak at block size of about 30 units corresponding to the prevailing size of aggregations of dominant species.

The common pattern analysis of *Artemisia vulgaris* and *Sinapis arvensis* (Fig. 7) by the T3lqv method shows distinct negative association at BS 8 (64 cm), which corresponds to the size of morphological clumps of *Artemisia*. Clearly, *Sinapis* is competitively excluded from those clumps.

Discussion

The analysis of both artificial and real data suggests that the Pqv method gives relatively poor results. In particular, if there were several scales of aggregation, it was able to detect the smallest one only. Good results were obtained for the Ttlqv and particularly T3lqv methods. Surprisingly, Hill (1973) suggested the T3lqv method but found it to perform poorly, which is probably why it was not included in some subsequent comparisons (e.g. Zahl 1977, Ludwig and Goodal 1978, Ludwig 1979, Carpenter and Chaney 1983). The results of these comparisons differ according to the type of data used and according to criteria used to evaluate methods. For example, Carpenter and Chaney (1983) preferred the paired- to blocked-quadrats methods for detecting a single scale of aggregation. There

79

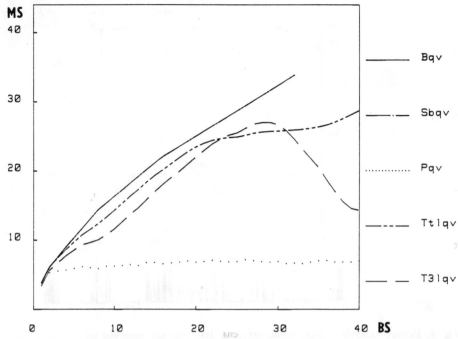

Fig. 6. Pattern analysis of the whole community. Only the five most important species are considered.

are two tasks included in analysis — description of the pattern (detection of scales of aggregation) and statistical testing. Whereas the Ttlqv and T3lqv methods have the best descriptive power, they do not allow rigid statistical testing. On the other hand, valid statistical tests are available for the Bqv method (Mead 1974) and for paired-quadrats method. However, the question remains as to which null hypothesis should be tested. The case where individuals are spaced randomly is very rare – testing the patently unrealistic hypothesis is not useful. It seems that use of various methods is preferable; no single method gives a complete insight into the various scales of aggregation (Usher 1975, 1983).

The type of data used is another important question. The original method was designed for density. The frequency used in this paper is less laborious, but also less precise and particularly in herbaceous vegetation the counting of individual tillers of vegetatively propagating organisms would be extremely time consuming. The information lost is mainly that concerning the pattern on the smallest scale — the morphological pattern. I argue that morphological pattern may be more efficiently studied directly (e.g. direct measurement of tussock diameter, examination of the belowground connections of particular ramets, etc.). The type of data should be kept in mind for interpretation of results — the scale of aggregation corresponds to the size of aboveground patches. It is usually considerably larger than the area where the patch is rooting.

In any case, pattern analysis is the first step and should be followed by testing of mechanisms.

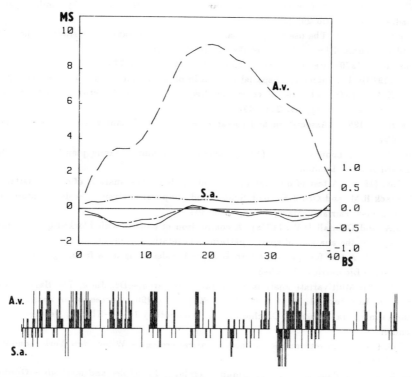

Fig. 7. Common pattern analysis of *Artemisia vulgaris* (A.v.) and *Sinapis arvensis* (S.a.).
— COMS, — · — · - rMS.

Conclusions

The following conclusions can be drawn from the analyses of both real and artificial data:

1. The most efficient method is the three term local quadrat variance (T3lqv). The blocked quadrat methods are more efficient than the paired quadrat methods.
2. The common analysis of two populations by the T3lqv method provided ecologically interpretable results about the spatial relationships of populations.
3. The pattern of the whole community may be described by a method using squared Euclidean distances between adjacent blocks.

References

Carpenter S.R. and Chaney J.E. (1983): Scale of spatial pattern: four methods compared.— Vegetatio 53: 153–160.

Galiano E.F. (1983): Detection of multi-species patterns in plant populations.—Vegetatio 53: 129–138.

Gibson D.J. and Greig-Smith P. (1986): Community pattern analysis: a method for quantifying community structure.— Vegetatio 66: 41–47.

Goodall D.W. (1974): A new method for the analysis of spatial pattern by random pairing of quadrats.—Vegetatio 29: 135–146.

Greig-Smith P. (1952): The use of random and contiguous quadrats in the study of the structure of plant communities.— Annals of Botany 16: 293–316.

Greig-Smith P. (1979): Pattern in vegetation.—J. Ecol. 67: 755–779.

Hill M.O. (1973): The intensity of spatial pattern in plant communities.—J. Ecol. 61: 225–235.

Kershaw K.A. (1957): The use of cover and frequency in the detection of pattern in plant communities.—Ecology 38: 291–299.

Kershaw K.A. (1961): Association and co-variance analysis of plant communities.—J.Ecol. 49: 643–654.

Kershaw K.A. and Looney J.H.H. (1985): Quantitative and dynamic plant ecology.—3rd ed. Edward Arnold, London.

Ludwig J.A. (1979): A test of quadrat variance methods for the analysis of spatial pattern.—In: Cormack R.M. and Ord J.K.[eds.], Spatial and temporal analysis in ecology. Intern. Co-op. Publ. House, Fairland, Maryland, pp. 289–304.

Ludwig J.A. and Goodall D.W. (1978): A comparison of paired- with blocked-quadrat variance methods for the analysis of spatial pattern.—Vegetatio 38: 49–59.

Mead R. (1974): A test for spatial pattern at several scales using data from a grid of continuous quadrats.—Biometrics 30: 295–307.

Orlóci L. (1978): Multivariate analysis in vegetation research.—Dr. Junk, The Hague, Boston.

Osbornová J., Kovářová M., Lepš J. and Prach K. [eds.] (1990): Succession in abandoned fields. Studies in central Bohemia, Czechoslovakia. (Geobotany 15). —Kluwer, Dordrecht, Boston, London.

Pielou E.C. (1969): An introduction to mathematical ecology.—Wiley, New York, London; Sydney, Toronto.

Pielou E.C. (1974): Population and community ecology. Principles and methods.—Gordon and Breach, New York, Paris, London.

Ripley B. (1987): Spatial point pattern analysis in ecology.—In: Legendre P. and Legendre L. [eds.], Developments in numerical ecology. Springer, Berlin etc., pp. 407–429.

Usher M.B. (1969): The relation between mean square and block size in the analysis of similar patterns.—J. Ecol. 57: 505–514.

Usher M.B. (1975): Analysis of pattern in real and artificial plant populations.—J. Ecol. 63: 569–586.

Usher M.B. (1983): Pattern in the moss-turf communities of the Sub-Antarctic and maritime Antarctic.—J. Ecol. 71: 945–958.

Zahl S. (1977): A comparison of three methods for the analysis of spatial pattern.—Biometrics 33: 681–692.

Krahulec F., Agnew A.D.Q., Agnew S. & Willems J.H. [eds.]: *Spatial processes in plant communities.* pp. 83–88.

Multi-scaled and multi-species pattern analysis in boreal forest communities

Alexandr A. Maslov

Laboratory of Forest Science, Academy of Sciences of the USSR, Uspenskoe, Moscow region, 143030 USSR

Keywords: Axes interpretation, Ecological indicator values, Groups of associated species, Multi-scaled pattern analysis, Principal components analysis.

Abstract. Multivariate pattern analyses were carried out in the nine forest communities of the Moscow region (USSR). Vegetation was sampled in parallel transects 40–70 m long. Each transect is a linear sequence of 20×20 cm quadrats. Scales of spatial heterogeneity (multi-species spatial patterns) were defined by means of principal components analysis and nested blocks combined application. In different communities the results reveal the existence of 3–6 scales. Groups of associated species were chosen at each of the scales. Ellenberg's and Landolt's ecological "indicator values" were used as an aid in the interpretation of ordination diagrams in ecological context. In most cases axes I and II are interpretable as a gradient of structural (edificator) influence of tree species. The role of particular factors at different spatial scales is discussed. A conclusion has been drawn that multi-species intracoenotic patterns must be studied at different spatial scales and the role of different factors depends strictly on the scale used.

Introduction

A multi-scaled and multi-species pattern analysis has been, of late, one of the most rapidly growing areas of plant ecology (Bouxin and Gautier 1982, Maslov 1983, Galiano 1983, Gibson and Greig-Smith 1986, Zobel and Masing 1987). This approach hinges on the thesis that vegetation is a multi-level phenomenon, while the scale of pattern ranges from the region of influence of an individual plant to the geographical (Greig-Smith 1979). Obviously, each scale has intrinsic peculiarities of spatial structure and temporal changes and, by virtue of this fact, must be studied separately.

For a number of years, the author has studied the horizontal structure of boreal forest communities in the Moscow region (USSR). The purpose of the present paper was to detect scales and identify the causes of pattern of particular communities at the beginning of permanent observations (monitoring). The objectives of the study were: (a) to detect the typical sizes of multi-species spatial patterns; (b) to describe groups of associated species, and (c) to identify causes of pattern at different spatial scales.

83

Study area and sampling

The study area belongs to the subzone of coniferous-broadleaved forests. All nine undisturbed natural forest communities were studied in forest reserves. The predominant tree species are *Pinus sylvestris* and *Picea abies*. The age of the investigated stands varies from 55 to 100 years. The following types of forest were studied: (1) *Vaccinium*-type pine forest (*V. vitis-idaea*), (2) *Myrtillus*-type pine forest (*Vaccinium myrtillus*), (3) *Polytrichum*-type pine forest, (4) *Eriophorum*-type pine forest, (5) *Oxalis*-type spruce forest, (6) *Oxalis*-type spruce forest with *Populus tremula*, (7) *Galeobdolon*-type spruce forest with *Tilia cordata*, (8) *Myrtillus*-type spruce forest and (9) *Sphagnum*-type spruce forest. For a more detailed description of forest types see appropriate reference (Maslov 1989b).

Permanent sample plots of 0.1 to 0.28ha were located subjectively in relatively homogeneous stands. Vegetation of the ground layer was sampled in permanent parallel transects, 40 to 70m long. Each transect was a linear sequence of 20 × 20 cm quadrats. The total of such quadrats per sample plot ranged from 1000 to 4000. Rooted frequency was determined for all species of terricolous mosses and lichens, herbs and dwarf shrubs. Species with frequencies ranging from 1 to 90% were selected for further study; the local frequency in each nested block was determined.

Analysis and results

To quantify mean sizes of multi-species patterns, we used a combination of nested blocks and principal components analysis (Maslov 1983). We assumed that the amount of variance contributed by the first three axes adequately reflects the extent of all species' association, just as covariance reflects association between two species (Kershaw 1960). In the absence of patterns of mean sizes, the contributions of the first few axes to the overall variance should steadily grow with increasing block size, as variance will decline. However, plateaux or small peaks in the axes-contribution-versus-block-size graphs (at least, on one of the axes) can be interpreted as the levels of spatial heterogeneity. The proposed method makes it possible to quantify not just individual multi-species patterns along the transect, but generalized scales or levels of spatial heterogeneity. It should be noted that an approach similar in concept was suggested earlier by Noy-Meir and Anderson (1970).

Detection of spatial heterogeneity scales in a particular community can be seen in the case of the *Eriophorum*-type pine forest (Fig. 1). The 0.6m and 1m scales of pattern correspond here to the size of nano- and micro-relief elements, and the 6m scale to aggregate sizes of some individual species. In all, the number of scales detected in different forest types ranged from 2 to 5 or 7. Detection of scales in each community enabled us to use strictly specified block sizes in subsequent work on discriminating groups of associated species and identifying the causes of pattern.

Species were grouped according to their loadings on the first two principal components (R-type with varimax rotation). It was demonstrated successfully that in different communities results of ordination at different levels may have a widely varying degree of differentiation. For example, in the case of a *Myrtillus*-type pine forest, ordination diagrams at all levels were found to be very much alike. Presumably, the causes of pattern at different spatial scales in that community are the same. Contrary to this, in the *Vaccinium*-type and the *Eriophorum*-type pine forests, as well as in the *Oxalis*-type mixed

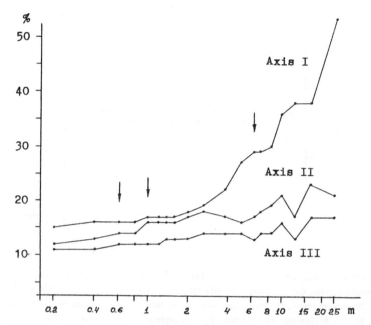

Fig. 1. Plots of variance (%) against block size for the first three components in the *Eriophorum*-type pine forest. The scales of spatial heterogeneity are indicated by arrows.

forest, results of ordination are in large measure determined by the block size. On these grounds, we assumed that in these communities the particular causes and their role are not the same at different levels.

In Fig. 2, one can see ordination diagrams in the *Vaccinium*-type pine forest at the scales of 1m and 7m. The first diagram is a good reflection of actual phytocoenotic interactions between species. The groups of species in the second diagram depict, essentially, the stages of dynamics in the community, where regeneration succession follows a ground fire (*Cladina* species → *Luzula pilosa* → *Vaccinium vitis-idaea* → *V. myrtillus*).

Axes interpretation in ecological terms was carried out, using Ellenberg's (1979) and Landolt's (1977) ecological indicator values. At first (Maslov 1985) Persson's (1981) approach was employed. Later (Maslov 1988) it was shown that in different communities the same species may occupy a different position along the environmental factor gradient. On this basis we suggest a new, more accurate method of interpreting ordination axes with the aid of indicator values. First, indicator values for light, moisture, acidity, nitrogen, and humus supply were calculated for each 20 × 20 cm quadrat by averaging the indicator values of all the species present. Then, given these data, improved estimates for species indicator values were calculated using weighted averaging method. The latter technique is similar to that of direct gradient analysis. For details, see Persson (1981) and ter Braak and Gremmen (1987). As a result, to interpret the axes, we use niche characteristics (weighted averages) obtained for each community rather than individual indicator values.

85

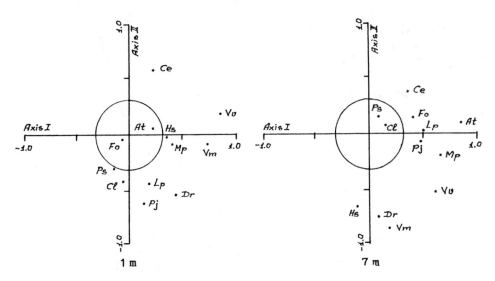

Fig. 2. Ordination diagrams at the scales 1m and 7m in the *Vaccinium*-type pine forest. At — *Agrostis tenuis*, Ce — *Calamagrostis epigeios*, Cl — *Cladina arbuscula + C. rangiferina*, Dr — *Dicranum rugosum*, Fo — *Festuca ovina*, Hs — *Hylocomium splendens*, Lp — *Luzula pilosa*, Mp — *Melampyrum pratense*, Pj — *Polytrichum juniperinum*, Ps — *Pinus sylvestris* juv., Vm — *Vaccinium myrtillus*, Vv — *Vaccinium vitis-idaea*.

Table 1. Spearman rank correlation coefficients significant at $p < 0.10$ (*) and at $p < 0.05$ (**). Axes interpretation in ecological terms using weighted averages in *Oxalis*-type spruce forest with *Populus tremula*. L — light, M — moisture, R — reaction, N — nitrogen supply, H — humus supply.

Block size	Axis	L	M	R	N	H
1.6 m	1	-0.06	0.34	0.43*	0.66**	-0.18
	2	0.60**	0.71**	-0.04	0.43*	0.46**
3.6 m	1	-0.01	0.33	0.07	0.59**	0.04
	2	0.51**	0.72**	0.03	0.48**	0.46**
5.0 m	1	-0.01	0.35	0.33	0.61**	-0.08
	2	0.46**	0.26	-0.21	-0.16	0.43*
10.0 m	1	-0.01	0.26	0.09	0.55**	-0.02
	2	-0.42*	0.07	0.36	-0.31	-0.02

Table 1 shows results of axes intepretation by ecological factors in the *Oxalis*-type mixed forest. One can see distinctly that a major cause of pattern here is the supply of

Table 2. Spearman rank correlation coefficients significant at $p < 0.10$ (*) and at $p < 0.05$ (**). Axes interpretation in phytocoenotic terms using weighted averages in *Galeobdolon*-type spruce forest with *Tilia cordata*. For explanation see text.

Block size	Axis	*Picea abies*	*Quercus robur*	*Tilia cordata*	Young *Picea abies*	Young *Tilia cordata*
1.0 m	1	0.90**	0.55**	0.41**	-0.24	0.45**
	2	0.28*	0.26*	0.34*	-0.41**	0.23
1.4 m	1	0.32	0.66**	0.36*	-0.12	0.51**
	2	-0.17	0.30	0.25*	-0.34*	0.39**
2.0 m	1	0.40**	0.68**	0.47**	-0.42**	0.34*
	2	0.34*	0.38*	0.58**	-0.59**	0.15
5.0 m	1	0.31*	0.69**	0.46**	-0.37**	0.28
	2	-0.26	0.65**	0.54**	-0.60**	0.29

available nitrogen in the soil. At all scales of spatial heterogeneity the first, and in some cases the second axis are correlated with nitrogen. Moisture comes as a secondary cause, its role being very important only with scales up to 5m, which is due to the size of windfall pits and mounds. Next in order of importance come light flux and humus supply.

The procedure of detection of phytocoenotic factors of pattern comprised two stages (Maslov 1989a). At the first stage we studied ground layer species affinities with respect to single tree species, by performing a direct ordination of species along the gradients that reflected the edificator influence of tree species. Intensity of the influence on the nearest tree quadrat was measured using different formulae; for example, $F = r^2/l^2$, where F = force of influence, r = tree stem radius, l = distance from tree stem centre to quadrat centre. Later, on the strength of these data, a weighted average was determined for each species.

It appears from the calculation of weighted averages that ground layer species in all communities gravitate to or shy away from particular tree species, although the number of such species with such a pronounced trend is always small. For instance, in the *Galeobdolon*-type mixed forest, at 1m scale, groups of species are related not to the differences in environment formed by coniferous *Picea abies* or broadleaved *Quercus robur, Tilia cordata*, but to a much simpler gradient "gaps — closed tree canopy" (see Table 2). Nonetheless, correlations between the axes and tree species at all the scales are close, and at the 5m scale the differences between *Tilia cordata* and *Quercus robur* and young growth of *Picea abies* become most prominent. At the same time, we have managed to show that with most communities the influence of tree species is effected by modification of environmental characteristics.

The findings of our study prove that there are several levels (scales) of spatial heterogeneity in all communities. This clearly calls for quadrats different in size, the optimal alternative being their "adjustment" to average size of respective pattern scales. A combination of this principle with methods of objective (statistical) detection of horizontal structural elements is, in my opinion, a guarantee of successful development of the theory of plant cover heterogeneity.

References

Bouxin G. and Gautier N. (1982): Pattern analysis in Belgian lime-stone grasslands.—Vegetatio 49: 65–83.

Ellenberg H. (1979): Zeigerwerte der Gefässpflanzen Mitteleuropas. 2 Aufl.—Scripta Geobot. 9: 1–122.

Galiano E.F. (1983): Detection of multi-species patterns in plant populations.—Vegetatio 53: 129–138.

Gibson D.J. and Greig-Smith P. (1986): Community pattern analysis: a method for quantifying community mosaic structure.—Vegetatio 66: 41–47.

Greig-Smith P. (1979): Pattern in vegetation.—J. Ecol. 67: 755–779.

Kershaw K.A. (1960): The detection of pattern and association.—J. Ecol. 48: 233–242.

Landolt E. (1977): Ökologische Zeigerwerte zur Schweizer Flora.—Veröff. Geobot. Inst. ETH 64: 1–208.

Maslov A.A. (1983): On combined application of the block method and the principal components analysis in the study of the mosaic structure of forest communities. 1. Axes determination.— Bull. Mosk. Ob. Isp. Prir., Ot. Biol., 88(6): 73–79. [Russ.]

Maslov A.A. (1985): On combined application of the block method and the principal components analysis in the study of the mosaic structure of forest communities. 2. Axes interpretation in ecological terms.—Bull. Mosk. Ob. Isp. Prir., Ot. Biol., 90(4): 107–117. [Russ.]

Maslov A.A. (1988): On estimation of forest plants niche characteristics by use of ecological indicator values.—In: Kull K. and Ksenofontova T. [eds.], Perspectives of the theory of phytocoenology, Tartu, pp. 105–110. [Russ.]

Maslov A.A. (1989a): On combined application of the block method and thge principal components analysis in the study of the mosaic structure of forest communities. 3. Axes interpretation in phytocoenotic terms.—Bull. Mosk. Ob. Isp. Prir., Ot. Biol., 94(3): 89–95. [Russ.]

Maslov A.A. (1989b): Small-scale pattern of forest plants and environmental heterogeneity.— Vegetatio 84: 1–7.

Noy-Meir I. et Anderson D.J. (1970): Multiple pattern analysis, or multiscale ordination: towards a vegetation hologramm?—In: Statistical Ecology 3: 207–225.

Persson S. (1981): Ecological indicator values as an aid of the interpretation of ordination diagrams.—J. Ecol. 69: 71–84.

ter Braak C.J.F. and Gremmen N.J.M. (1987): Ecological amplitudes of plant species and the internal consistency of Ellenberg's indicator values for moisture.—Vegetatio 69: 79–87.

Zobel M. et Masing V. (1987): Bogs changing in time and space.—Arch. Hydrobiol., Beih. Ergebn. Limnol., 27: 41–55.

Krahulec F., Agnew A.D.Q., Agnew S., Willems J.H. [eds.]: *Spatial processes in plant communities.* pp. 89–98.

Treefall disturbance — a factor structuring vegetation in boreal spruce forests

Bengt Gunnar Jonsson

Department of Ecological Botany, University of Umeå, S-901 87 Umeå, Sweden

Keywords: Diversity management, Regeneration, Treefall gaps, Windthrow.

Abstract. Plant communities are dynamic and heterogeneous on many spatial and temporal scales. Natural disturbance is one factor creating this heterogeneity. It is important to recognize the components of the disturbance regime, as they affect both community patterns and life histories of individual species. In forest ecosystems one common natural disturbance is the uprooting of trees. Some of the spatial and temporal patterns created by treefalls in boreal forests are discussed. Wind is the primary agent uprooting trees, but soil factors, topography and biotic factors also contribute. It is possible to find both spatial and temporal aggregation of treefalls at different scales. Uprooting strongly influences the vegetation in boreal forests: both the tree layer by providing regeneration sites; the ground layer by maintaining high bryophyte diversity. The forest areas needed for a natural pattern of uprootings could be used as criteria for the size of nature reserves and as a basis for reserve management.

Introduction

Earlier ideas of community organization were strongly influenced by the general assumption that equilibrium situations prevail in most communities, i.e. population sizes fluctuate around stable mean values. The basis for this assumption is that resources are at an almost constant level in both time and space. However, resource availability is very variable in many cases (Fig. 1). This has led authors (e.g. Pickett 1980, Wiens 1984) to argue that in most communities non-equilibrium conditions dominate. The recognition of temporal and spatial scales is very important. If we study an ecosystem for a sufficiently long time we will always notice changes, due to processes like species migration, glaciation, climatic changes and species evolution. On the other hand, on a very small spatial scale we will notice change due to factors such as small scale disturbance and random establishment. At intermediate scales it might be possible to find equilibrium conditions. This implies that we have to define the temporal and spatial scales of our studies carefully before we consider equilibrium or non-equilibrium conditions.

We can state, therefore, that all communities are dynamic and spatially heterogeneous, at least on some spatial or temporal scale (White 1979, Sousa 1984). One of the most important processes creating such heterogeneity is natural disturbance. In this paper I

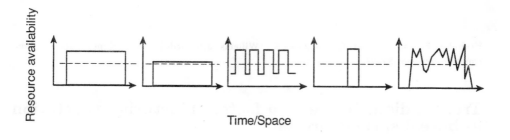

Temporal aspect	Constant, high	Constant, low	Seasonal	Ephemeral	Unpredictable
Spatial aspect	Continous, high	Continous, low	Patchy	Isolated	Unpredictable

Fig. 1. Resource availability patterns in time and space. The dotted line represents lowest resource level needed for reproduction. (Modified after Southwood 1977 and Keddy 1989.)

will consider disturbance as "a discrete, punctuated killing, displacement, or damaging of one or more individuals (or colonies) that directly or indirectly creates opportunity for new individuals (or colonies) to become established" (Sousa 1984).

The resource availability pattern (Fig. 1) can be regarded as the template for life-history evolution (Southwood 1977). It should be important, therefore, to determine attributes of the disturbance regime, as it is one of the mechanisms affecting resource availability. The general descriptors of the disturbance regime, (i) areal extent, (ii) magnitude, (iii) frequency, (iv) predictability and (v) rotation period (Sousa 1984) are dependent on the actual agent of disturbance, and the effect on population and community structure is thus linked to this agent.

In the boreal forests a number of disturbance regimes operate concurrently, mainly fire, animal activities and wind. Fires have a large impact on the tree species composition (Larsen 1980), age structure (Zackrisson 1977), and regeneration (Mirov 1967). Animal activities, such as vole grazing, runways, carcasses, digging activities and droppings from large herbivores, can create both favourable regeneration sites (Ericson 1977) and micro-habitats (Cameron and Wyatt 1986) for a number of species.

Strong winds cause tree mortality resulting in both canopy gaps and forest floor disturbance. Catastrophic windthrows may parallel the effects of fire, for instance by creating regular regeneration waves as described from high-altitude balsam fir forest (Sprugel and Bormann 1981). However, even on a smaller local scale, gap-phase dynamics due to single treefalls may have a large impact on the community structure (Shugart 1984, Hytteborn *et al.* 1987).

In this paper I will discuss the temporal and spatial pattern of uprooted trees in boreal spruce forests and some of the implications for the vegetation. Most of the data discussed was obtained on research conducted by a research group from the University of Umeå. We have recently developed a method for dating single treefalls and have, therefore, a system which allows us to study the disturbance history and different aspects of small scale successions. This work is ongoing, and at present (1989) we are studying the temporal and spatial patterns of treefalls in two virgin spruce forests (Kirjesålandet and Gardfjället)

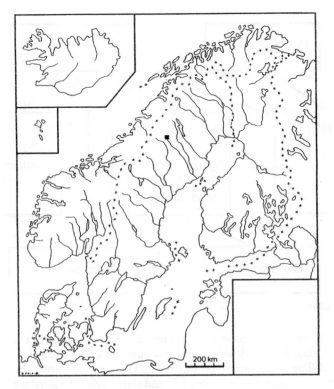

Fig. 2. Scandinavia, with the research area indicated by a square.

in Västerbotten, north Sweden (Fig. 2).

Determinants of uprooting in boreal spruce forests

Several factors may contribute to the uprooting of a tree (Fig. 3) and it may be difficult to determine a single causal factor.

Strong winds are the most obvious factor involved in treefall. The importance of high wind speeds can be seen by comparing wind data with the age of dated treefalls in the same area. Data from the spruce forest at Kirjesålandet shows that the period between 1970–1973 was characterized by both hard and frequent storms which resulted in numerous treefalls (Fig. 4). The relationship between duration of the storm and treefall is, however, unclear, but it seems reasonable that long periods with high winds will weaken the root system by repeated movement of trees (Bosshard 1967, Stone 1977).

In Scandinavia most storms are westerly in origin and this is reflected in the direction of treefalls (Fig. 5). In Kirjesålandet, with 1–150 years old treefalls, the fall direction was more or less the same as for treefalls during a single autumn in the nearby forest, Gardfjället (Fig. 5).

Local factors, mainly topography, may strongly influence the risk of trees being wind damaged. In regions with rough topography smaller areas are affected by stormfellings

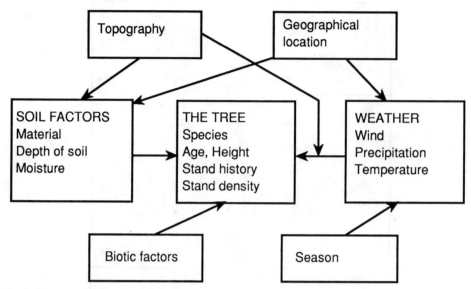

Fig. 3. Factors affecting treefall formation.

(Persson 1975). However, trees on steep slopes are often tilted and therefore easily uprooted (Schaetzl *et al.* 1989). Forests close to open areas such as mires and clearcuttings are more subject to stormfelling as high wind speeds can build up over open areas (Sernander 1936). Other factors of importance are heavy rainfalls and snowloads associated with storms. The most extensive stormfellings recorded in Scandinavia occurred during years with heavy snowloads (Persson 1975).

Trees on wet, gleyed or organic soils are generally more subject to uprooting than trees on other soils, especially if the ground water table is high (Schaetzl *et al.* 1989). Such conditions affect the growth form of the roots, resulting in an extensive but shallow and badly anchored root system (Fraser 1962). However, trees on moraine with large boulders can also be rather loosely anchored and are easily uprooted (Sernander 1936). These examples illustrate that anything preventing deep rooting makes trees susceptible to uprooting (Schaetzl *et al.* 1989).

Different tree species vary in susceptibility to uprooting (Putz *et al.* 1983, Naka 1982). This is probably a result of differences in the overall shape of the tree and the rooting pattern (Schaetzl *et al.* 1989), but factors such as tree density (Putz *et al.* 1983) and stem form (Larsen 1963) may also contribute. In Scandinavian forests, there does not seem to be any clear difference between the two dominant conifers (*Picea abies* and *Pinus sylvestris*). Instead they seem develop the root systems in relation to soil type and water availability (Persson 1975).

It has been shown that larger (Putz 1983) and older trees (Dunn *et al.* 1983) are more sensitive to uprooting than smaller and younger ones. In the study in the spruce forest on Gardfjället larger trees were more exposed to wind induced mortality than smaller trees. In addition, breaks were more common than uprooting among the larger trees. In the

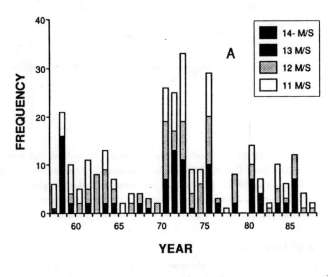

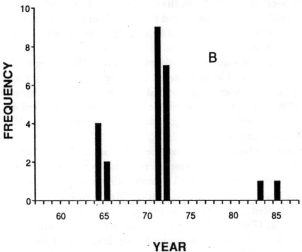

Fig. 4. A. Number of three hour periods with wind speeds over 10 m/s during the period 1957–1987 at the meteorological station at Gunnarn 90 km from Kirjesålandet. (Data from the Swedish meteorological and hydrological institute.). B. Number of uprooted trees in 2.7 ha of Kirjesålandet during the period 1957–1987.

Kirjesålandet study a majority (85%) of the windfalls were breaks. The high proportion of breaks is probably a result of high frequency of heart rot in large old trees.

93

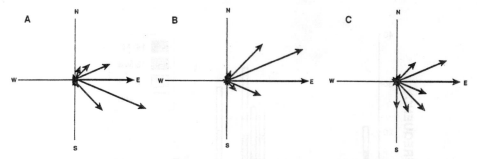

Fig. 5. A. Wind vectors (1957–1987) for wind speeds exceeding 12 m/s at the meteorological station at Gunnarn 90km from Gardfjället and Kirjesålandet. The summed length of the vectors in each figure equals to 100%. (Data from the Swedish meteorological and hydrological institute.). B. Fall-direction for treefalls at Gardfjället felled during one autumn (1986; n=338). C. Fall-direction for uprooted trees of 1–150 years age in Kirjesålandet (n=96).

Spatial distribution of treefalls

A treefall can either be composed by a single tree, a "simple fall", or by two or more trees, a "compound fall" (Schaetzl *et al.* 1989). In the studies from Gardfjället and Kirjesålandet 42% and 51% of the windfalls were "compound falls", indicating that when one large tree blows down, other trees are often affected and felled, either by becoming more exposed to the wind or by physically being knocked down.

In Kirjesålandet the distribution of treefalls was significantly aggregated (Fig. 6).. The aggregation is probably an effect of local patterns in topography, mainly areas with wetter ground and areas with large boulders. For the Nature Reserve "Fiby urskog" in southeast Sweden, Sernander (1936) published maps of the spatial pattern created by two storms (1931 and 1932) which also showed local aggregation of treefalls at a scale which largely coincides with that of Kirjesålandet.

At present there seem to be only a few studies of regional differencies in distribution of treefalls. General topography and location in relation to mountain chains may result in completely different treefall frequencies (Alexander 1964). However, it has been noted that more trees become uprooted at higher altitudes (MacKenzie 1976, Booth 1977).

The size of the uprooted patch depends on the size of the trees involved and the shape of their root systems (Putz 1983). In Kirjesålandet with natural treefall processes the maximum size was 9.0 m². In the experimentally fragmented forest at Gardfjället (stand sizes range between 1/16 – 1 ha situated on a large clearcutting) the uprootings were larger, maximum 21 m², suggesting that also the actual location of the forest stand, in relation to edges, influences the size of the uprooting.

In Kirjesålandet and Gardfjället there were 17 and 22 relatively young (1 to c. 200 years) treefalls/ha respectively. Similar studies from other boreal areas are lacking. The total disturbed area was approximately 0.3% of the forest floor in Kirjesålandet (Jonsson and Esseen, unpubl.). Other studies have shown values ranging from 0.09–50% of the forest floor composed of pits and mounds (Schaetzl *et al.* 1989). The definition of pits and mounds varied, however, between these studies, thus hampering comparison.

94

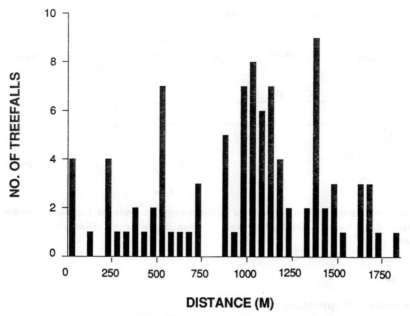

Fig. 6. Number of uprootings in 50×30m parts of a 1850 m long transect in Kirjesålandet.

Temporal distribution of treefalls

Treefalls are not evenly distributed over the seasons. In Scandinavia, most treefalls occur during October to February when soil water levels are high (Persson 1975), and are associated with high wind speeds. During the other parts of the year there are fewer storms, less water in the soil and/or ground frost.

As mentioned above, certain storms can result in large numbers of treefalls (cf. Fig. 5). A notable example is the catastrophic storm in May 1795, in the province of Uppland, Sweden, where almost 25 cbm/ha were windthrown in a 17 500 ha area (Sernander 1936). In Kirjesålandet 67% of the treefalls between 1957 and 1987 were felled in 1971–1973. In the Nature Reserve "Fiby urskog" there seem to have been approximately five large storms per century that could be traced in the structure of the forest (Sernander 1936).

Regeneration waves observed in balsam fir (*Abies balsamea*) forest (Sprugel and Bormann 1981) may have parallels in other forests. During some extensive storms both before 1964 and during the period 1975–1978, no treefalls were observed in Kirjesålandet. This may indicate that more or less cyclic patterns may exist. As in the case of forest fires, biomass accumulates and trees get older increasing the susceptibility to both fire and catastrophic windthrow. Aggregation in certain diameter classes could, therefore, be a result of cyclic releases of suppressed trees. It should also be noted that wind-induced mortality is more size and age dependent than the mortality caused by fire. Estimates for the disturbance cycle of treefalls regarding time needed to "affect all parts of a stand" range from 100–25 000 years, with most estimates around a couple of thousand years (Schaetzl *et al.*

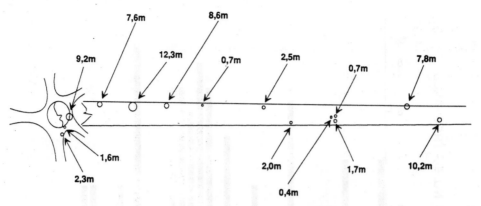

Fig. 7. Spruces regenerated on a single log in Blekaskogen in southern Lappland, Sweden. The height of the trees (indicating tree age) is given in the figure. (Redrawn after Arnborg 1942.)

1989).

Consequences of uprooting in boreal forest

Treefalls provide a number of favourable regeneration sites for trees (Sernander 1936, Arnborg 1942; Fig. 7). In a study from Blekaskogen, in southern Lappland, Sweden, the majority of saplings had regenerated on spruce logs, stumps and in pits and mounds (Arnborg 1942). Elsewhere on the forest floor the thick moss and humus layer prevent establishment of spruce (Sirén 1955, Arnborg 1942). Treefalls also create canopy gaps where the suppressed saplings can reach the canopy layer (Sernander 1936, Foster and Reiners 1986).

The abundant fallen logs in virgin spruce forests are a very important substrate for many rare bryophytes (Söderström 1983). Compared with managed forest stands, there are more and larger logs and the bryophyte flora is much richer in the virgin stand (Söderström 1988).

The effects of uprooting on the forest floor bryophyte community were studied in Kirjesålandet (Jonsson and Esseen, unpubl.). Patches where the forest floor had been disturbed had twice as many species compared with undisturbed forest floor (112 and 56 respectively). Species richness was highest for treefalls of intermediate ages, but the vegetation composition (in terms of dominance and presence of early successional species) were different from that of the surrounding forest floor for at least a period of 100 years. Four main causes of the observed diversity pattern were proposed: i) uprooting creates space, free from potential competitors, that is available for bryophyte colonization; ii) disturbed patches have high habitat heterogeneity; iii) within-patch disturbance continues long after patch formation through erosion from the tip-up mound; iv) the presence of an extensive diasporbank (Jonsson, unpubl.) and the small patch size implies a short distance to potential sources of bryophyte diaspores which should increase the chance of establishment.

Conclusion

The dynamics of forest floor vegetation created by treefalls are of great ecological importance in forested ecosystems, and the spatial and temporal scales on which they operate require further study. Not only the number of windfalls per unit area, but also the size- and age-distributions should be taken into account as these influence the bryophyte vegetation of the treefall. To conclude, the forest area needed for a natural treefall pattern could be applied as one important criterion when discussing the size of Nature reserves. In the case of existing reserves the possibility of deliberately creating treefalls should be considered.

Acknowledgements. I thank M. Dynesius, P.-A. Esseen and J. Anderson for valuable comments on the manuscript. The work has been financially supported by the Hierta-Retzius fund for scientific research.

References

Alexander R.R. (1964): Minimizing windfall around clearcuttings in spruce-fir forests.—For. Sci. 10: 130–142.

Arnborg T. (1942): Lågaföryngringen i en sydlappländsk granurskog.—Svenska Skogsvårdsföreningens Tidskrift 40: 47–78.

Booth T.C. (1977): Windthrow hazard classification.—Forestry Comm. (U.K.) Res. Inf. Not. No. 22-77-SILN.

Bosshard W. (1967): Erhebungen über die Schäden der Wintersturme 1967.—Schw. Z. Forstw. 118: 806–820.

Cameron R.G. and Wyatt R. (1986): Substrate restriction in.enthomophilous *Splachnaceae*: Role of spore dispersal.—Bryologist 89: 279–284.

Dunn C.P., Gutensperger G.R. and Dorney J.R. (1983): Catastrophic wind disturbance in an old-growth hemlock-hardwood forest, Wisconsin.—Can. J. Bot. 61: 211–217.

Ericson L. (1977): The influence of voles and lemmings on the vegetation in a coniferous forest during a 4-year period in northern Sweden.—Wahlenbergia 4: 1–115.

Foster J.R. and Reiners W.A. (1986): Size distribution and expansion of canopy gaps in a northern Appalachian spruce-fir forest.—Vegetatio 68: 109–114.

Fraser A.I. (1962): The soil and roots as factors in tree stability.—Forestry 35: 117–127.

Hytteborn H., Packham J.R. and Verwijst T. (1987): Tree population dynamics, stand structure and species composition in the montane virgin spruce forest of Vallibäcken, north Sweden.—Vegetatio 72: 3–19.

Keddy P.A. (1989): Competition.—Chapman & Hall, London.

Larsen J.A. (1980): The boreal ecosystem.—Academic Press, New York.

Larson P.R. (1963): Stem form development of forest trees.—For. Sci. Monogr. 5: 1–42.

MacKenzie R.F. (1976): Silviculture and management in relation to risk of windthrow in Northern Ireland.—Ir. For. 33: 29–38.

Mirov N.T. (1967): The genus *Pinus*.—The Roland Press Company, New York.

Naka K. (1982): Community dynamics of evergreen broadleaf forest in southwestern Japan. I. Wind damaged trees and canopy gaps in evergreen oak forest.—Bot. Mag. (Tokyo) 95: 385–399.

Persson P. (1975): Stormskador på skog. Uppkomstbetingelser och inverkan av skogliga åtgärder.—Dept. Forest Yield Research, Research note 36.

Pickett S.T.A. (1980): Non-equilibrium coexistence of plants.–Bull. Torrey Bot. Club 107: 238–248.

Putz F.E. (1983): Treefall pits and mounds, buried seeds, and the importance of soil disturbance to pioneer trees on Barro Colorado Island, Panama.—Ecology 64: 1069–1074.

Putz F.E., Coley P.D., Lu K., Montalvo A. and Aiello A. (1983): Uprooting and snapping of trees: structural determinants and ecological consequences.—Can. J. For. Res. 13: 1011–1020.

Schaetzl R.J., Johnson D.L., Burns S.F. and Small T.W. (1989): Tree uprooting: review of terminology, process and environmental implications.—Can. J. For. Res.19: 1–11.

Sernander R. (1936): Granskär och Fiby urskog. En studie över stormluckornas och marbuskarnas betydelse i den svenska granskogens regeneration.—Acta Phytogeog. Suec. 8: 1–232

Shugart H.H. (1984): A theory of forest dynamics. The ecological implications of forest succession models.—Springer-Verlag, New York.

Sirén G. (1955): The development of spruce forest on raw humus sites in northeast Finland and its ecology.—Acta For. Fenn. 62: 1–408.

Söderström L. (1983): Hotade och sällsynta arter i norrländska granskogar.—Svensk Bot. Tidskr. 77: 4–12.

Söderström L. (1988): The occurrence of epixylic bryophyte and lichen species in an old natural and a managed forest stand in northeast Sweden.—Biol. Conserv. 45: 169–178.

Sousa W.P. (1984): The role of disturbance in natural communities.—Ann. Rev. Ecol. & Syst. 15: 353–391.

Southwood T.R.E. (1977): Habitat the template for ecological strategies.—J. Animal Ecol. 46: 337–365.

Sprugel D.G. and Bormann F.H. (1981) Natural disturbance and the steady state in high altitude balsam fir forest.—Science 211: 390–393.

Stone E.L. (1977): Abrasion of tree roots by rock during wind stress.—For. Sci. 23: 333–336.

White P.S. (1979): Pattern, process and natural disturbance in vegetation.—Bot. Rev. 45: 229–297.

Wiens J.A. (1984) On understanding a non-equilibrium world: Myth and reality.—In: Strong D.R., Simberloff D. and Abele L.G. [eds.], Ecological communities: conceptual issue and the evidence.—Princeton, New Jersey, pp. 439–457.

Zackrisson O. (1977): Influence of forest fires on North Swedish boreal forest.—Oikos 29: 22–32.

Krahulec F., Agnew A.D.Q., Agnew S., Willems J.H. [eds.]: *Spatial processes in plant communities.* pp. 99–109.

Dispersal and distribution patterns in patchy, temporary habitats

Lars Söderström

Department of Ecological Botany, University of Umeå, S-901 87 Umeå, Sweden.

Keywords: Distribution, Dispersal, Patch dynamics, Regional dynamics, Epixylic bryophytes

Abstract. In patchy habitats, rarity and commonness can be viewed in several ways. Usually a species is regarded as rare when it occurs in few localities in an area. Another way of viewing rarity is to regard it in connection with the number of suitable localities. A species will then be rare if it occurs in only a minority of the suitable localities, and common if it occurs in all suitable localities, in a region even if there are only a few localities available. Within localities, the populations of species will be large or small. Combining the regional and local occurrence gives four distribution patterns. *Core* species occur abundantly in the majority of suitable localities, *urban* species abundantly in only a few of the suitable localities, *rural* species unabundantly in the majority of suitable localities and *satellite* species unabundantly in a few of the suitable localities. The application of this model (the CURS model) is discussed with examples from bryophytes. Dispersal ability and population dynamics are regarded as the main factors creating these distribution patterns.

Introduction

Distribution patterns and rarity of species are often a good starting point for asking questions about communities and the processes regulating the abundance and distribution of species. Whether a species is rare or common depends on the definition of rarity and commonness. Several ways of viewing rarity have been presented. On a global (or at least regional) scale, a species may be rare only in certain localities (diffusive rarity) or rare all over its range (suffusive rarity; Schoener 1987). Rabinowitz (1981) distinguishes three variables along which a species may be rare: limited geographic ranges, narrow habitat requirements and small populations. All combinations of these are possible. She regards suffusive rarity as 'genuine rarity' and diffusive rarity as 'pseudo-rarity' and does not incorporate the last group in her seven forms of rarity. Certainly, suffusively rare species may more often be those designated endangered. However, the vast majority of the rare species in a given locality are common in other parts of their ranges. In the Swedish list of rare and endangered forest bryophytes (Ingelög *et al.* 1987) more than 85% of species have larger populations elsewhere (cf. Schoener 1987 for comparative figures on birds).

For diffusively rare species, rarity is a matter of spillover from places where the species is more common and an understanding of rarity actually involves an understanding of the dispersal power of the species, i.e. how it diffuses across the landscape.

The most common way of treating rarity and commonness is to estimate the size and number of populations in a region (e.g. Drury 1974, Adamus and Clough 1978, Brown 1984), e.g. in a country like Sweden or a mountain range like the Alps. It is commonly observed that a species occurs most abundantly in the center of its range and that the abundance will gradually decline towards the geographical limits (e.g. Delcourt *et al.* 1981, Hengeveld and Haeck 1982). Brown (1984) has presented a theory involving centers of many favourable localities for the species where the species is common. The density of localities is reduced further away from such a center and the species becomes progressively rarer. Hanski (1982) adopted another view of commonness and rarity when he compared actual and potential occupancies of possible localities. He observed two patterns in regional distribution which he described as the core and satellite species hypothesis. Some species occur abundantly in most of the available localities (core species) while others occur in only a few of the available localities, and only in low abundances (satellite species). Hanski states that a species rarely occurs sparsely on a majority of available localities or abundantly on few localities and points out the importance of immigration and emigration. When defining available substrate and localities for species it is of the greatest importance to know what specific demands each species has. Both the abiotic (e.g. moisture) and biotic (e.g. competition) factors must be considered. The demands during the whole life cycle must be incorporated since crucial stages (such as germination) may define which localities are available. It is difficult to clarify all demands of a species during the whole life cycle. In the theoretical discussion of this paper, I assume that the demands of species are fully understood and the bryophytes used as examples have been studied intensively during several years (Söderström 1987a,b, 1988a,b). Factors still unknown in their life cycle may account for some deviations from the hypothesis.

A result of treating the species in relation to available substrate is that a species occurring in few places with small populations, and generally treated as rare, may be a core species if available substrate is rare and the species grows on every spot with a suitable substrate. On the other hand, a generalist species growing in many localities in fairly large numbers and which is able to grow on a large number of substrates, is a satellite species if it is unable to colonize the majority of suitable localities.

Söderström (1989) detected four patterns when treating some epixylic (growing on decaying logs) bryophytes in the same way, i.e. abundant on a majority of available localities (core species), abundant but only in a few localities (urban species), sparse in a majority of available localities (rural species) and sparse in only a few localities (satellite species). Almost all abundantly occurring species could be found on suboptimal substrates indicating that dispersal within localities and establishment were good. But distance dispersal was postulated as the main factor restricting some abundant species to a few localities. From these observations, a theoretical model for distribution patterns is developed, the CURS model.

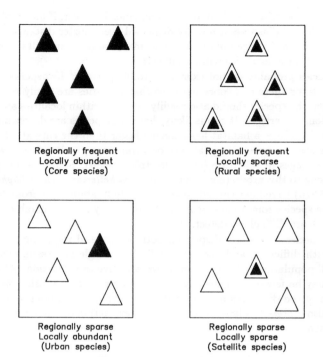

Fig. 1. Schematic patterns of distribution for core, urban, rural, and satellite species, respectively. Filled triangles are occupied.

The CURS model

The CURS model describes distribution patterns among species on temporary, patchy habitats but may be extended to all species in patchy habitats. The model is based on the pattern of occurrence in two different scales, abundance (frequency) within suitable, colonized localities and frequency of suitable localities occupied. The combination of these two scales gives four theoretical patterns (Fig. 1). Species occurring abundantly in a majority of localities correspond to Hanski's (1982) core species and species occurring sparsely on only very few suitable localities correspond to his satellite species. The other two combinations, abundant only in a few of the suitable localities and sparse in a majority of suitable localities, are called urban and rural species, respectively (Söderström 1989). These patterns may be explained by differences in dispersal ability and population stability.

Core species can occur if the distances between localities are shorter than the efficient dispersal distance of the species. In these cases, the contribution of diaspores from outside reduces the effects of a temporal shortage in locally produced diaspores. The more stable a populations the less important is diaspore contribution from outside.

Urban species exist when the environment is coarse-grained compared with the dispersal ability of the species. If diaspores are poorly dispersed over longer distances, popu-

101

lations never or rarely receive diaspores from outside. However, if diaspore production (or growth) is great, and diaspores are easily dispersed over shorter distances, the species can flourish locally. Even if the populations are normally large, some will become extinct and recolonization of these localities will be difficult.

Rural species probably do not exist as a stable pattern. If dispersal is efficient over longer distances, or there is a spore bank, so that localities are easily recolonized after a local extinction, the species should also readily disperse within localities and build up large local populations. Therefore, it seems likely that rural species are decreasing or increasing in population size where a habitat has changed faster than the rate at which the population dynamics has had time to regulate the occurrences, e.g. when forests are fragmented, resulting in shortages in substrate (cf. Söderström 1988b). Some species previously considered to belong to this group (e.g. *Anastrophyllum hellerianum* and *Lophozia ascendens*, Söderström 1987b) have later turned out to fluctuate in abundance from year to year and appear as core species some years. It might be that they survive harsh periods as resting diaspores which are difficult to detect.

Satellite species are poorly dispersed, both locally and regionally, or become established only with difficulty, and occur sporadically. Single successful colonizations may result in small populations which, however, are sensitive to extinction. At the edge of the range there may be few and irregularly occurring events of establishment so that only a small fraction of the diaspores are successful. Occurrences are often temporal. Satellite species may also represent a late stage in regional extinction or early stages in an invasion or a range extension.

CURS patterns arising from different dispersal abilities

A few examples will illustrate the importance of dispersal ability in species and the impact of locality density. Bryophytes occurring on decaying logs in spruce forests are patchily distributed on two scales, logs within a forest and forests within the landscape. I have investigated the occurrence of 18 species confined to decaying logs and have been able to detect at least three of the groups (Söderström 1989; Fig. 2). The suitability of logs determined for each species separately, resulting in different amounts of suitable substrates for different species at the same locality.

Two of the core species, *Ptilidium pulcherrimum* and *Tetraphis pellucida*, produce large diaspores (spores and gemmae, resp.) which presumably do not disperse very easily (cf. Crum 1972, Mogensen 1983, van Zanten and Pócs 1981). At least the former species does not fluctuate very much between years (Jonsson and Söderström 1988, Söderström unpubl.). Therefore, they appear to rely on population stability, while the other core species rely on efficient dispersal since they produce small diaspores in large quantities. They also vary considerably in population size and population structure between years (Söderström, unpubl.).

The urban species regularly produced gemmae, which are rather large, but they were never recorded with sexual reproduction. Most of them occurred regularly on suboptimal substrates, indicating efficient local dispersal and establishment (cf. Söderström 1987a). Thus, it appears that they have no problem of dispersal within the locality but some restrictions in dispersal over longer distances.

102

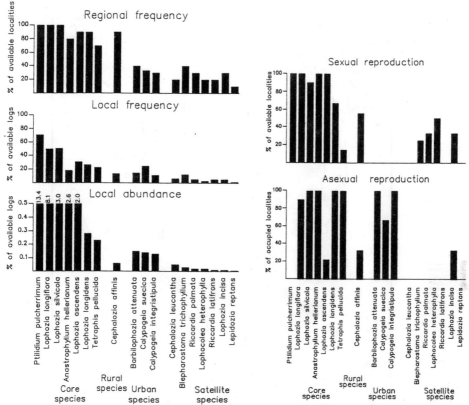

Fig. 2. Frequency and abundance at local and regional scales and frequency of reproduction in 18 bryophyte species categorized as core, urban, rural and satellite species. Numbers on top of bars under local abundance denote % of available logs when the values are > 0.5% (from Söderström 1989).

The satellite species rarely produced any diaspores and none of the species was found more than once with sexual reproduction. The small populations, together with the rarity of reproduction, result in very few diaspores available for dispersal.

CURS patterns arising from different density of localities

The impact of density of localities on the distribution patterns may be illustrated by the occurrence of *Orthodontium lineare*, a recent invader, in southern Sweden (Hedenäs *et al.* 1989). On a regional scale, *O. lineare* shows a marked southwestern distribution, occurring in many localities in northwestern Skåne and southern Halland and only in a few localities in Västergötland. It is absent in the eastern and central parts of southern Sweden. As the prevailing winds in southern Sweden are southwesterly, one would expect a species like *O. lineare* to have spread also to areas further north and east. Instead, the regional pattern is correlated with the density of suitable forests, whereas climatic factors

103

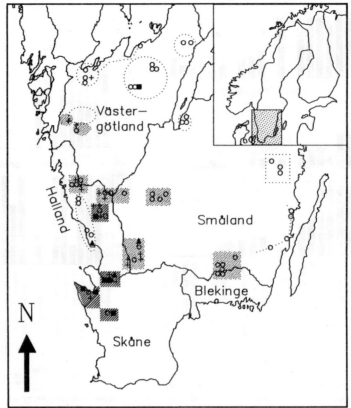

Fig. 3. Occurrence of *Orthodontium lineare* in relation to substrate amount. Hatched area—
region with many suitable localities, dotted area—region with intermediate density of suitable localities, enclosed within dotted line—region with few suitable localities, dotted line—
travelled distance with few localities. Filled square—large populations, filled triangle—
intermediate populations, plus—small populations, open circle—absent (after Hedenäs *et al.*
1989).

seem to be of less importance (Fig. 3; Hedenäs *et al.* 1989).

Where the density of suitable localities is high, localities where *O. lineare* has disappeared may be efficiently recolonized from nearby localities and *O. lineare* will occur as a
core species. This is the case in northern Skåne and southern Halland, where large parts
of the natural broadleaved forests have been cut down and replaced with conifer forests,
mostly of spruce, which are very suitable for *O. lineare*. In Småland and Västergötland,
conifer forests are more natural and the broadleaved forests remain largely unreplaced.
In these forests, the man-made disturbances are not so great and available localities are
more scattered. Therefore, the chances of colonization are much smaller, and *O. lineare*
will occur only in a few of the localities, i.e. as a satellite species. Where localities are

very scattered, as in the interior of Västergötland, single colonizations may be possible and the species may build up large local populations if substrate is abundant. But as the localities are so far apart from each other, dispersal between them is restricted and the species occurs as an urban species.

Discussion

The more temporary a substrate is the more often a species has to move around and the greater is the importance of good dispersal ability for the survival of a population. Small populations are more sensitive to stochastic extinctions than large ones (MacArthur and Wilson 1967, Gilpin and Soulé 1986) and species with such small, extinction-sensitive, populations have to disperse to new places more often than species with large populations if the total population size is not to decrease. Small populations also produce fewer diaspores and each diaspore must be more efficient in order to balance the smaller number. The greater the distance between localities, the more efficient must be the transport of diaspores between localities.

If the local extinction rate is high and dispersal is important, three questions may be asked. (1) How often must a species be successfully dispersed to balance local extinctions? (2) How large are the distances between localities? (3) How efficient is dispersal ability of the species? Both environmental and population variables influence the dispersal. Important environmental variables are patch size, duration of substrates and pattern (grain size) of the environment. Among the population variables, duration of populations and dispersal ability are important. The dispersal ability, i.e. how often a diaspore will establish in a new locality, depends on (1) the number of diaspores produced, (2) how efficiently diaspores are transported to the new locality and (3) how easily diaspores become established in a new locality.

The proportion of produced spores establishing new populations is low among bryophytes. In *Ptilidium pulcherrimum* only 0.6% of the spores produced are deposited on a suitable substrate within the locality (Söderström and Jonsson 1989). Not all of these are established and the proportion of diaspores lost during transport between localities must be considerable. Assuming that these losses are of the same proportion as during transport within localities, not more than 1/20,000 of the produced spores will establish new populations in a new locality. The losses during transport between localities are certainly much greater than within localities, probably several magnitudes larger.

Three groups of rarity have already been distinguished by Drury (1974), few in a few places, few in many places and many in few places. The fourth group, many in many places, are usually considered common and not treated by Drury (cf. also Adamus and Clough 1978). These groups are superficially the same as my groups but Drury does not relate occurrence to the availability of substrate. Instead, he uses restricted demands as one explanation for the occurrence of his groups. Nor does he use difference in dispersal ability as an explanation for his model. Species occurring as few in a few places (superficially my satellite species) are considered to be species restricted to rare habitats. In my sense, these may still be core species if they are able to colonize most such places. Drury also states that he has no explanation as to the pattern of many in a few places (superficially my urban species). He says that it is doubted that "a set of narrow niches exists widely spaced

105

geographically and yet placed as if available for a suitable adapted species to occupy them". This pattern is, however, not rare in environments which are fragmented into smaller and well separated localities, e.g. forests in agricultural areas (cf. Curtis 1956)

Species are usually rare near geographical distribution limits because the available localities are fewer (e.g. Brown 1984). In northern Sweden, *Lophocolea heterophylla* is found on some of the northernmost localities (Arnell 1956, Söderström 1983). It is a satellite species there, but occurs more frequently on available localities in central and southern Sweden. Occasionally successful establishments in the north give small populations. However, the very rare occurrence of diaspore production there prevents establishment of large populations and dispersal to new areas. In southern Sweden, the spore production is large and *L. heterophylla* occurs almost everywhere although no studies of the occurrence in relation to available substrate have been done. On the other hand, common species such as *Barbilophozia attenuata* and *Calypogeia integristipula*, are urban species and thus regionally rare. An explanation of this may be that they have a wide niche space and may grow in a large number of localities but are not able to colonize every available locality.

Environments are changing rapidly nowadays, some decreasing and others increasing in area. Fragmentation of environment has three effects on dispersal opportunities. First, the individual localities become smaller and the amount of substrates decreases. If the number of new substrates created per unit of area is low, situations with substrate shortage may occur and the risk of extinction will increase. Second, the local population size decreases and the populations produce fewer diaspores and are even more prone to become extinct. Third, the distance between localities increases which reduces the number of diaspores transported between localities.

If the local populations become so small, and the distances between localities so large that the increasing extinction rate is not compensated by distance-transported or locally produced diaspores, core species will decrease and convert either to an urban species or continue towards extinction through a stage of rural species (or urban species?) and a stage of satellite species (Fig. 4). Decreasing species may survive locally as long as the substrates remain but some populations will, by chance, become extinct.

Populations of urban species can disappear in, e.g. an environmental catastrophe and never recolonize. Some species can be so stable and depend on so few events of diaspore immigration that the only effect of a fragmentation on the species will be disappearence of some populations together with their localities. On the other hand, an increase in the environment area may decrease the distances between suitable localities to less than the efficient dispersal distance and the species may increase.

Among epixylic bryophytes in northern Sweden, two species, *Anastrophyllum hellerianum* and *Lophozia ascendens*, generally regarded as rare and threatened forest species in Sweden (Ingelög *et al.* 1987, Söderström 1983), are core species (Söderström 1989). Rare core species are very restricted and narrow in their demands but are able to colonize most of the available localities. They can, therefore, serve as indicator species for specific habitat conditions in patchy habitats. *A. hellerianum* and *L. ascendens* are both indicative of moist, late successional forests with large logs in intermediate decay stages (cf. Söderström 1988a). Urban species are, on the other hand, not very suitable as indicator species since they will be absent from a large number of the suitable localities. However, they are often

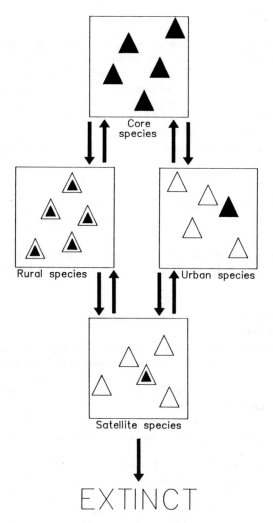

Fig. 4. Possible pathways of species changing patterns between core, rural, urban and satellite species when their suitable environment is fragmented (downward arrows) or increasing in area (upward arrows). Filled triangles are occupied.

used as arguments for nature conservancy as they occur abundantly but locally (many in a few places *sensu* Drury 1974). This is often interpreted as evidence of an exceptional locality (e.g. Adamus and Clough 1978). However, a study of the population dynamics and dispersal ability of the species might result in a different interpretation.

Concluding remarks

The regional and local frequencies (and local abundance) are correlated. In most cases, a large local population produces large numbers of diaspores and the absolute number of diaspores successfully dispersed is as a rule larger than in small populations. The number of localities occupied will then be large. If the number of occupied localities is large (i.e. in regionally frequent species) the total population size in the region will be higher and the probability of successful recolonization of localities will increase. Therefore, most of the species should be either core or satellite species (as Hanski, 1982, pointed out). The urban (and rural) patterns may be regarded as exceptions arising when dispersal is less efficient. A simulation model (Herben et al., in prep.) also indicates that dispersal-limited systems are rarer than habitat-limited systems. But the more variable the establishment ability and the more randomly the patches are arranged, the more dispersal-limited the system will be. The spore transport and the distance between localities seem to be the most important factors for persistence of species in temporary habitats. A closer study of reproduction and dispersal should increase the understanding of regional dynamics in plants.

Acknowledgements. I thank Lars Ericson and Christer Nilsson for valuable comments on the manuscript and Lars Hübinette for correcting the English text.

References

Adamus P. L. and Clough G. C. (1978): Evaluating species for protection in natural areas.—Biol. Conserv. 13:165–178.

Arnell S. (1956): Illustrated Moss Flora of Fennoscandia. I. Hepaticae.—Lund.

Brown J. H. (1984): On the relationship between abundance and distribution of species.—Am. Nat. 124:255–279.

Crum H. (1972): The geographical origin of the mosses of North America's eastern deciduous forest.—J. Hattori Bot. Lab. 35:269–298.

Curtis J. T. (1956): The modification of mid-latitude grasslands and forests by man.—In: Thomas W.L. [ed.], Man's role in changing the face of the earth. Univ. of Chicago Press. pp. 721–736.

Delcourt H. R., West D. C. and Delcourt P. A. (1981): Forests of the south eastern United States: quantitative maps for aboveground woody biomass, carbon, and dominance of major tree taxa.—Ecology 62: 879–887.

Drury W. H. (1974): Rare species.—Biol. Conserv. 6:162–169.

Gilpin M. E. and Soulé M. E. ((1986): Minimum viable populations: processes of species extinction.—In: Soulé M. [ed.], Conservation Biology. Sinauer, Massachusetts. pp. 19–34

Hanski I. (1982): Dynamics of regional distribution: the core and satellite species hypothesis.—Oikos 38: 210–221.

Hedenäs L., Herben T., Rydin H. and Söderström L. (1989): Ecology of the invading moss species Orthodontium lineare in Sweden: Spatial distribution and population structure.—Holarctic Ecology 12: 163–172.

Hengeveld R. and Haeck J. (1982): The distribution of abundance. I. Measurements.—J. Biogeogr. 9: 303–316

Ingelög T., Thor G. and Gustafsson L. (1987): Floravård i skogsbruket.—Artdel. Skogsstyrelsen, Jönköping. 2nd ed.

Jonsson B.G. and Söderström L. (1988): Growth and reproduction in the leafy hepatic *Ptilidium pulcherrimum* (G. Webb.) Vainio during a 4 year period.—J. Bryol.15: 315–325.

MacArthur R. H. and Wilson E. O. (1967): The theory of island biogeography.—Monographs in population biology 1. Princeton.

Mogensen G. (1983): The spore.—In: Schuster R. M. (ed.), New manual of bryology. Hattori Bot. Lab., Nichinan, pp. 325–342.

Rabinowitz D. (1981): Seven forms of rarity.—In: Synge H. (ed.), The biological aspects of rare plant conservation. Chichester.

Schoener T. W. (1987): The geographical distribution of rarity.—Oecologia 74:161–173.

Söderström L. (1983): Hotade och sällsynta mossarter i norrländska granskogar.—Svensk Bot. Tidskr. 77:4–12.

Söderström L. (1987a): Dispersal as a limiting factor for distribution among epixylic bryophytes.—Symp. Biol. Hung. 35:475–484.

Söderström L. (1987b): The regulation of abundance and distribution patterns of bryophyte species on decaying logs in spruce forests.—Ph. D. thesis, Univ. of Umeå, Umeå.

Söderström L. (1988a): Sequence of bryophytes and lichens in relation to substrate variables of decaying coniferous wood in northern Sweden.—Nord. J. Bot. 8: 89–97.

Söderström L. (1988b) The occurrence of epixylic bryophyte and lichen species in an old natural and a managed forest stand in northeast Sweden.—Biol. Conserv. 45: 169–178.

Söderström L. (1989): Regional distribution among bryophyte species on a temporal, patchy substrate.—Bryologist 92: 349–355.

Söderström L. and Jonsson B. G.. (1989): Spatial distribution and dispersal in the leafy hepatic Ptilidium pulcherrimum.—J. Bryol. 15:793–802.

van Zanten B. O. and Pócs T. (1981): Distribution and dispersal of bryophytes.—Adv. Bryol. 1: 479–562.

Krahulec F., Agnew A.D.Q., Agnew S. & Willems J.H. [eds.]: *Spatial processes in plant communities.* pp. 111–120.

Sapling establishment patterns in relation to light gaps in the canopy of two primeval pine-spruce forests in Sweden

Rik Leemans

International Institute of Applied Systems Analysis, A-2361 Laxenburg, Austria. Present address: *National Institute of Public Health and Environmental Protection, P.O. Box 1, 3720 BA, Bilthoven, The Netherlands.*

Keywords: Canopy structure, Forest dynamics, Gap dynamics, Spatial pattern.

Abstract. The spatial pattern of seedlings, saplings and canopy trees was studied in two primeval forests in central Sweden. Canopy and forest structure was determined in three 0.25 ha plots. Life stage classes were distinguished on the basis of size distributions. Ripley's K-function (1977) combined with Monte Carlo simulation was used to analyze the spatial pattern within each class. Saplings and sub-canopy trees were strongly aggregated and canopy trees were randomly dispersed. The proportion of individuals growing in gaps was used to relate the spatial patterns of saplings and sub-canopy trees with the occurrence of small canopy gaps. Monte Carlo simulation, using fixed sapling positions and randomly repositioned canopy gaps, confirmed the importance of canopy gaps for spruce regeneration. The appearance of understorey trees in small gaps suggest that gaps are necessary for the successful establishment of spruce saplings and that gaps are more often closed by recruitment of new individuals into the sapling bank, rather than by the release of suppressed trees.

Introduction

Canopy gaps, created by the death of a tree, drive forest dynamics (Platt and Strong 1989). Differences in gap sizes result in specific regeneration patterns of species, based on their shade tolerance characteristics (Canham 1989; Whitmore 1989). In tropical forests, small gaps release suppressed saplings of climax species, while in large gaps pioneer species can also establish (Swaine and Whitmore 1988). This pattern is probably similar for most forests of the world (Whitmore 1982).

The dynamics most commonly identified are the so-called gap-phase dynamics (Watt 1947) in which a large mature tree becomes large enough to dominate its environment. After its death a gap is formed in which shade-intolerant, early successional species become established and thus resulting in a cyclic successional pathway. Gap-phase dynamics are observed in many tropical forests (Brokaw 1985; Denslow 1980; Lawton and Putz 1988) and temperate forest (e.g. Bray 1956; Clebsch and Busing 1989; Ishikawa and Ito 1989; Runkle 1985a; White *et al.* 1985).

111

The dynamics which Sernander (1936) describes for boreal forests in high latitudes differ from gap-phase dynamics in that shade-intolerant species cannot establish after the formation of small gaps. This is due to the small lateral crown extensions of individual mature trees. The death of such a tree creates too small a gap to provide light intensities at the forest floor sufficient for shade-intolerant species to establish. Succession in these forests is typically directional and proceeds from shade-intolerant species towards dominance of shade-tolerant species (Bergeron and Dubuc 1989).

Recruitment of new canopy trees into gaps take place from newly germinated seedlings as well as from elongated growth of suppressed saplings (Sernander 1936). Shade-intolerant species are only capable of establishing successfully after a large disturbance in which many canopy trees are removed from the canopy and large gaps are created. Large scale disturbances as generated by fire (Bergeron and Dubuc 1989; Bonan and Shugart 1989; Foster 1985; Heinselman 1981) and windthrow (Hytteborn and Packham 1987; Webb 1989) are thus important for the maintenance of the shade-intolerant species. Sernander (1936) stated that shade-tolerant species regenerated successfully under a closed canopy and can grow very slowly for decades under a closed canopy. Height growth increases fast when a canopy gap is formed above by the death of one or more canopy trees. This regeneration strategy demands that suppressed saplings can be found randomly dispersed throughout the forest, in order to effectively utilize the higher light levels in newly created gaps. In this study I describe the forest and canopy structure of three boreal/boreal-nemoral stands in central Sweden (Sjörs 1965) and quantify the spatial patterns of different sized trees in relation to canopy gaps. Such patterns should demonstrate differences in the importance of a suppressed sapling bank or of direct regeneration and establishment in canopy gaps for shade-tolerant species.

Methods

Study areas

The study areas are the old-growth spruce forests of Fiby Urskog (59°54′ N, 17°22′ E) and Granskär (60°23′ N, 18°16′ E), in eastern central-Sweden. Both forests are relatively well known from the studies of Hesselman (1935), Sernander (1936), Hytteborn and Packham (1985), Hytteborn and Packham (1987) and Skarpe et al. (1989). The recent history of Fiby Urskog is essentially a succession, started after a major disturbance by a severe storm in 1796. Since this event the forest has also been relatively free from human impact. This has resulted in a valuable old-growth forest which is now a nature reserve. Granskär is located close to the coast and developed during a primary succession after its site emerged from the Baltic Sea 200-350 years ago (Skarpe et al. 1989). The dominant tree species in both areas is Norway spruce (*Picea abies* (L.) Karst.). Scots pine (*Pinus sylvestris* L.), aspen (*Populus tremula* L.) and birch (*Betula pubescens* Ehrh. and *Betula pendula* Roth.) are present but less abundant.

Both forests have many characteristic features of old-growth forests, with a mosaic of canopy gaps, decaying logs on the forest floor and dead standing trees. The understorey is poorly developed, and consists of a few scattered patches of hazel (*Corylus avellana* L.) and mountain ash (*Sorbus aucuparia* L.). Patches of spruce saplings of varying sizes are scattered throughout. Different types of field layer can be distinguished, depending on

112

local drainage and soil nutrient status: the nutrient-poor and well-drained soils developed over till are dominated by the dwarf shrubs bilberry (*Vaccinium myrtillus* L.) and cowberry (*V. vitis-idaea* L.) and many bryophyte species, whereas the more nutrient-rich soils on the marine clays have many small herbs and few shrubs. Intermediate stands are dominated by *Vaccinium myrtillus, V. vitis-idaea*, wood sorrel (*Oxalis acetosella* L.) and a few other herbs.

Sample plots

I established plots 1 and 2 in the spruce forest of Fiby Urskog in 1984, and plot 3 in Granskär in 1986. Each plot covered a square area of 0.25 ha (50 × 50 m). The plots were located so that the different spruce forest types were sampled. Plot 1 is on clay with some small patches of till. The dominant tree species is spruce. The herb layer consists of many different species. This stand was probably managed during the 19th century and early 20th century, but has been left to nature during the last decades. Plot 2 is on till with patches of clay. Spruce is the most abundant tree species, but pine, aspen and birch are also common. The herb layer consists mainly of bryophytes with patches of both *Vaccinium* species, and some small herbs such as *Oxalis*. Plot 3 in Granskär is similar to plot 2 in Fiby and was added to the analysis because of the differences in its natural history.

Sampling methods

I delineated all study plots and gridded them into 5 × 5 m squares. Within each square, I mapped all trees and saplings and recorded their vertical crown projections. Tree characteristics, such as diameter at breast height (DBH), tree height, bole height, crown diameter and crown depth were measured. I identified canopy gaps in all plots by determining their center and circumference and estimated their area by overlaying a 1 m^2 grid on the canopy maps and counting the grid squares occupied by a gap. I classified saplings as growing in a gap or under the canopy during the field studies.

Spatial analysis

I used Ripley's K-method (1977) to analyze the spatial patterns of individual trees. This method uses all distances between individual plants and is therefore superior to many of the nearest neighbor methods (e.g. Simberloff 1979) or distance distributions (Galiano 1982). $K(t)$ is defined as the expected number of individual trees within a distance t of a randomly chosen individual. The null hypothesis is of complete spatial randomness, i.e. a Poisson process, for which the expected value of $K(t)$ is πt^2. Edge effects are corrected (Diggle 1983; ter Braak 1980) and the results are transformed to $K^*(t) = K(t)/(\pi - t)$. This gives a function with an expectation of 0 for a randomly distributed pattern, positive values for an aggregated distribution and negative values for a regular distribution. A plot of $K^*(t)$ against t reveals spatial pattern at various scales. I used Monte Carlo simulations with random re-positioning of all individuals to determine confidence limits around the function.

To correlate canopy gaps with the observed spatial pattern of saplings and sub-canopy trees, the 'gap-sapling establishment ratio' (GSER) is defined. This is the ratio between the number of individuals growing under the canopy and of those growing in canopy gaps. Under the null hypothesis of independence the expected value for this statistic would equal

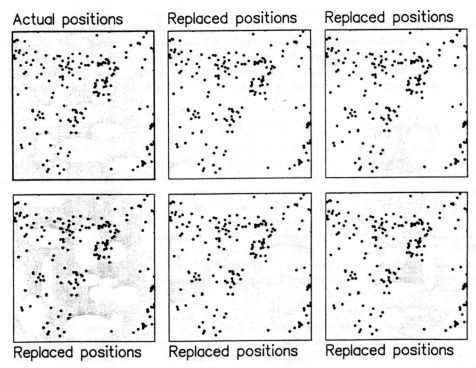

Actual positions   **Replaced positions**   **Replaced positions**

Replaced positions   **Replaced positions**   **Replaced positions**

Fig. 1. Canopy gaps described as ellipse of plot 3 in Granskär (above left) and 5 simulated canopies with repositioned gaps. The shape of the gaps are approximated by ellipses.

the canopy gap ratio (gap area to plot area) for the stand. To exclude edge effects, each plot was wrapped as a torus, giving a continuous area without borders (Ripley 1981). Significance levels around the expected value are determined by using Monte Carlo simulations with random re-positioning of canopy gaps. The significance levels of the GSER were computed by approximating the area and extent of a single gap as an ellipse (Runkle 1985b). The GSER was computed for both the observed canopy structure of the real plots and for 95 simulations with the repositioning of actual gaps (ellipses) outline (Fig 1.). This number of simulations defines a 99% confidence interval.

Results and Discussion

Stand structure and tree life-phases

The horizontal and vertical structures of the different plots are given in Figures 2–4. The bimodal height distributions seen in all plots are typical for old-growth forests dominated by shade-tolerant species (Shugart 1984). The tallest trees are 32 m, 27 m and 29 m respectively for plots 1, 2 and 3. Based on the size distributions one can classify all of the tree individuals into groups, representing different life phases. Phase 1 (saplings) includes all individuals between 0.5 m and 1.3 m (breast height). Phase 2 (sub-canopy trees) includes all trees larger than 1.3 m that occupy the lower parts of the continuous

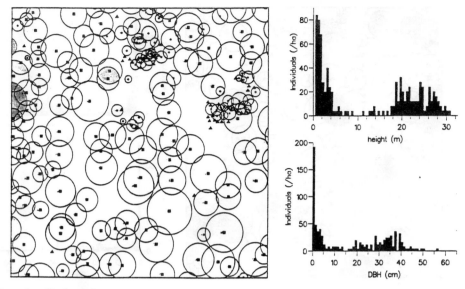

Fig. 2. Horizontal canopy structure and size distributions for all species in plot 1 (gray shade: *Picea abies*; light gray shade: *Pinus sylvestris*; waves: *Populus tremula*; grid: *Betula pubescens*; triangles: saplings; dots: sub-canopy trees; squares: canopy trees).

Table 1. Canopy gap characteristics in the 0.25ha plots in the spruce forests of Fiby Urskog and Granskär. The simulated area is the total gap area as computed by overlaying ellipses to represent the gaps.

	Plot 1	Plot 2	Plot 3
0– 50 m^2	1	2	3
50–100 m^2			2
100–150 m^2	1	2	3
150–200 m^2	2	1	1
200–250 m^2	1		
> 250 m^2	1	1	
total number (ha)	24	24	36
total area (m^2/ha)	1716	3332	3316
simulated area (m^2/ha)	1696	3282	3368

115

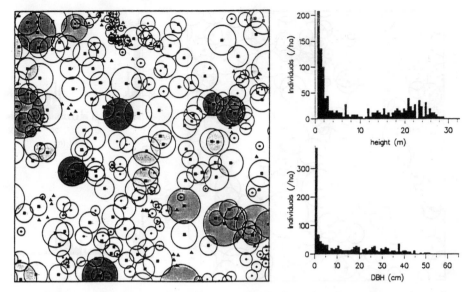

Fig. 3. Horizontal canopy structure and size distributions for all species in plot 2 (gray shade: *Picea abies*; light gray shade: *Pinus sylvestris*; waves: *Populus tremula*; grid: *Betula pubescens*; triangles: saplings; dots: sub-canopy trees; squares: canopy trees).

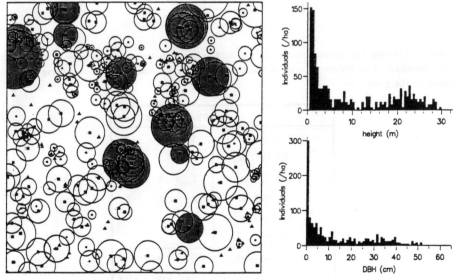

Fig. 4. Horizontal canopy structure and size distributions for all species in plot 3 (gray shade: *Picea abies*; light gray shade: *Pinus sylvestris*; waves: *Populus tremula*; grid: *Betula pubescens*; triangles: saplings; dots: sub-canopy trees; squares: canopy trees).

116

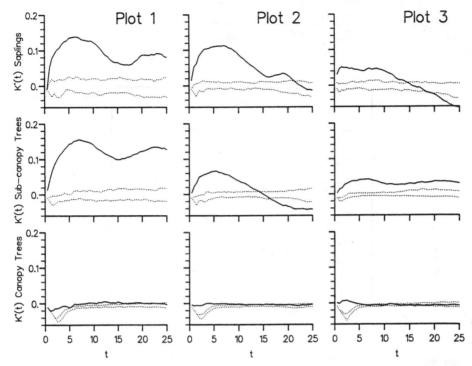

Fig. 5. $K^*(t)$ profiles (with Monte Carlo significance envelope) for the spruce saplings (B), sub-canopy trees (C) and canopy trees (D) (—, observed values; .... upper and lower 95% confidence interval).

forest canopy. Their upper height is set to 10 m. Phase 3 (canopy trees) forms the forest canopy and includes all individuals larger then 10 m. In the first three phases only spruce is represented. In the latter phases a few individuals of pine, aspen, birch, hazel and mountain ash are present.

Plot 1 has greater tree dimensions, basal area and crown diameters for spruce, but lower densities for pine, aspen and birch are less important here. Plots 2 and 3 have many small gaps between the crowns of single trees and this implies an open canopy structure. Canopy gap area (Table 1) due to the death or uprooting of one or more mature trees is largest in plots 2 and 3.

Structural differences between the plots are mainly explained by the site conditions of each plot. There is a higher soil moisture and nutrient availability in plot 1 as compared with plot 2 and 3, the latter probably having a more open canopy structure for this reason.

Spatial pattern in saplings and trees

Fig. 5 gives the $K^*(t)$ profiles for all life phases. Saplings and sub-canopy trees show a significant aggregation in all plots for small distances. At larger distances there are

117

Table 2. Relationships between small trees and canopy gaps, expressed by the gap-sapling establishment ratio (GSER). Simulated GSER is determined by replacing actual gap outlines by fitted ellipses. The confidence interval is for the null-hypothesis of no association between small trees and gaps.

	observed GSER	simulated GSER	99% confidence interval
Plot 1			
saplings	85.0	91.7	0.0–66.7
sub-canopy trees	82.1	71.8	0.0–43.6
Plot 2			
saplings	67.4	72.1	0.0–53.8
sub-canopy trees	77.0	72.1	0.0–47.5
Plot 3			
saplings	71.0	63.8	11.6–58.0
sub-canopy trees	66.4	65.0	19.7–54.0

major differences. Plot 1 displays the aggregated pattern throughout all distances, but plots 2 and 3 only up to distances of 10 m. The pattern then shifts through random to regular, becoming more regular with increasing distance. This regular pattern reflects the distribution of clusters of saplings. The change at 10 m is probably related to the canopy gap distribution. The average gap area is ca. 100 m² (Table 1) and many saplings are growing in these gaps. Sub-canopy trees showed highly significantly aggregated distributions (Fig. 5). A small decrease in the $K^*(t)$ value can be seen at ca. 10 m, but high values occur again at greater distances. An explanation for this pattern is that most of the sub-canopy trees grow in and directly around canopy gaps. This relates their observed pattern to the canopy gaps, but the relationship is not as strong as for the saplings. The $K^*(t)$ functions of canopy trees were not significantly different from a random distribution except for distances smaller than the average crown diameter.

Table 1 lists gap-sizes and total gap area for each plot. The more homogeneous canopy of plot 1 (Fig. 2) makes the forest stand less susceptible to storm damage, which may partly explain the lower gap frequency and gap area in this plot. Different stand histories magnify the differences in gap structure among plots. Plots 2 and 3 are from older stands with increased mortality rates and presumably higher susceptibility to windthrow, as is clearly displayed if the total numbers of dead trees are compared among all plots (Leemans 1986). Fewer trees have died during recent decades in plot 1, so there has been a higher rate of gap formation and/or enlargement in plots 2 and 3.

A relationship between gaps and the pattern of the smaller trees is apparent on inspection of maps (Fig. 2–4). Table 2 gives real and simulated GSER values with simulated 99%-confidence intervals for the null hypothesis of no relationship. In all plots the abundance of saplings proved to be significantly higher in gaps than under the forest canopy.

This result emphasizes the importance of gaps for the regeneration of spruce. The main qualitative difference between a canopy gap and the rest of the forest is the difference in light intensity (Nakashizuka 1985). The quantitative difference in light intensity is dependent on vegetation height, and on gap size and shape. Highest ratios were for plots 1 and 2, because of the more homogeneous closed canopy with consequently lower light levels beneath the canopy. The specific regeneration pattern was therefore more pronounced here than in plots 3–5.

Conclusion

By linking the observed spatial patterns of the different phases during the entire life cycle of the trees with the overall forest structure, this study has given insight into the processes driving the dynamics of the forest. Available light intensities within and beneath the forest canopy determine competition; this results in spatial patterns of tree individuals that are specific to the different life phases. Although spruce can tolerate low light intensity, successful establishment is related to the slightly increased light levels in canopy gaps. The dynamics of these old-growth boreal spruce forests are thus closely related to canopy gap formation and their successive enlargement during subsequent decades.

References

Bergeron Y. and Dubuc M. (1989): Succession in the southern part of the Canadian boreal forest.—Vegetatio 79: 51–63.

Bonan G.B. and Shugart H.H. (1989): Environmental factors and ecological processes in boreal forests.—Annual Review Ecol. Syst. 20: 1–28.

Bray J.R. (1956): Gap phase replacement in a maple-basswood forest.—Ecology 37: 598–600.

Brokaw N.V.L. (1985): Gap-phase regeneration in a tropical forest.—Ecology 66: 682–687.

Canham C.D. (1989): Different responses to gaps among shade-tolerant tree species.—Ecology 70: 548–550.

Clebsch E.E.C. and Busing R.T. (1989): Secondary succession, gap dynamics, and community structure in a southern Appalachian cove forest.—Ecology 70: 728–735.

Denslow J.S. (1980) Gap partitioning among tropical rain forest trees.—Biotropica 12 (supplement): 47–55.

Diggle P.J. (1983): Statistical analyses of spatial point patterns.—Academic Press, New York.

Foster D.R. (1985): Vegetation development following fire in *Picea mariana* (black spruce)—*Pleurozium* forests in southeastern Labrador, Canada.—J. Ecology 73: 517–534.

Galiano E.F. (1982): Pattern detection in plant populations through the analyses of plant to all plants distances.—Vegetatio 49: 39–43.

Heinselman M.L. (1981): Fire and succession in the conifer forest of northern North America.—In: West D.C., Shugart H.H. and Botkin D.B. [eds.], Forest succession: Concepts and application. Springer-Verlag, New York, pp. 374–405.

Hesselman H. (1935): Fibyskogen och dess utvecklungshistoria (Der Fibywald und seine Entwicklungsgeschichte).—Meddelanden från Stattliga Skogsförsöksanstalt 28: 525–570.

Hytteborn H. and Packham J.R. (1985): Left to nature: forest structure and regeneration in Fiby Urskog, Central Sweden.—Arboricultural Journal 9: 1–11.

Hytteborn H. and Packham J.R. (1987): Decay rate of *Picea abies* and the storm gap theory: a re-examination of Sernander plot III, Fiby urskog, Central Sweden.—Arboricultural Journal 11: 299–311.

Ishikawa Y. and Ito K. (1989): The regeneration process in a mixed forest in central Hokkaido, Japan.—Vegetatio 79: 75–84.

Lawton R.O. and Putz F. (1988): Natural disturbance and gap-phase regeneration in wind-exposed tropical cloud forest.— Ecology 69: 764–777.

Leemans R. (1986): Structure of the primaeval coniferous forest of Fiby.—In: Fanta J. [ed.], Forest dynamics research in Western and Central Europe. PUDOC, Wageningen, pp. 221–230.

Nakashizuka T. (1985): Diffused light conditions in canopy gaps in a beech (*Fagus crenata* Blume) forest.—Oecologia, Berlin, 66: 472–474.

Platt W.J. and Strong D.R. (1989): Gaps in forest ecology.—Ecology 70: 535–576.

Ripley B.D. (1977): Modelling spatial patterns.—Journal of the Royal Statistical Society, series B, 39: 172–212.

Ripley B.D. (1981): Spatial statistics.—John Wiley & Sons, Chichester, 252 pp.

Runkle J.R. (1985a): Disturbance regimes in temperate forests.—In: Pickett S.T.A. and White P.S. [eds.], The ecology of natural disturbance and patch dynamics. Academic Press Inc., Orlando, pp. 17–33.

Runkle J.R. (1985b): Comparison of methods for determining fraction of land area in treefall gaps.—Forest Science 31: 15–19.

Sernander R. (1936): Granskär och Fiby urskog, en studie över stormluckornas och marbuskarnas betydelse i den svenska granskogens regeneration. [The primitive forests of Granskär and Fiby).—Acta Phytogeographica Suecica 8: 1–232.

Shugart H.H. (1984): A theory of forest dynamics.—Springer-Verlag, New York, 278 pp.

Simberloff D. (1979): Nearest neighbor assessments of spatial configurations of circles rather than points.—Ecology 60: 679–685.

Sjörs H. (1965): Forest regions.—Acta Phytogeographica Suecica 50: 48–63.

Skarpe C., Hytteborn H. and Cramer W. (1989): Femtio års skogsutveckling påGranskär, norra Uppland.—Svensk Botanisk Tidskrift 83: 177–185.

Swaine M.D. and Whitmore T.C. (1988): On the definition of ecological species groups in tropical rain forest.—Vegetatio, 75: 81–86.

ter Braak C.J.F. (1980): Binary mosaics and point quadrat sampling in ecology.—IWIS-TNO A80 ST 9637, IWIS-TNO, P.O. Box 100, Wageningen, pp.

Watt A.S. (1947): Pattern and process in the plant community.— J. Ecol. 35: 1–22.

Webb S.L. (1989): Contrasting windstorm consequences in two forests, Itasca State Park, Minnesota.—Ecology 71: 1167–1180.

White P.S., MacKenzie M.D. and Busing R.T. (1985): Natural disturbance and gap phase dynamics in southern Appalachian spruce-fir forests.—Can. J. Forest. Res. 15: 233–240.

Whitmore T.C. (1982): On pattern and process in forests.—In: Newman E.I. [ed.], The plant community as a working mechanism. British Ecological Society, London, pp. 45–59.

Whitmore T.C. (1989): Canopy gaps and the two major groups of forest trees.—Ecology 70: 536–538.

Krahulec F., Agnew A.D.Q., Agnew S., Willems J.H. [eds.]: *Spatial processes in plant communities.* pp. 121–126.

Modelling pattern formation and its consequences in crops attacked by pests

Gareth Hughes

School of Agriculture, University of Edinburgh, Scotland, U.K.

Keywords: Wheat, *Delia coarctata*, Damage curves, Pest patches.

Abstract. Most pests and pathogens of plants have characteristically patchy spatial patterns, and thus can disrupt the more-or-less regular patterns in which most crops are sown, in addition to inflicting plant injury. However, most models of crop losses to pests and diseases deal only with the effect of intensity of attack and ignore the effect of spatial pattern of attack.

A model of crop loss to patchy pest attack is developed from the initial assumption that the rate of yield loss per unit plant injury depends on the spatial pattern of the attack. Spatial pattern is, in turn, expressed as a function of level of injury, providing a basis for the formulation of a critical point model in terms of the incomplete beta-function.

Previously collected data on the interaction between wheat (*Triticum aestivum*) crops and larvae of the wheat bulb fly (*Delia coarctata*) are used to illustrate some of the advantages and limitations of the model.

Introduction

Crops, being artificial populations of plants, have some spatial characteristics distinct from natural populations. They have definite boundaries, and are usually sown in regularly spaced rows, sometimes with regular spacing between the plants within rows. The between- and within-row spacings determine the average plant population density and rectangularity (a measure of spatial arrangement), both of which may influence yield per unit area (Willey and Heath 1969).

When a pest attack results in the death of plants, both the density and spatial arrangement of a crop are disrupted. Surviving plants are then, in many cases, able to respond to a relaxation in the level of intraspecific competition within the crop and to make some compensatory growth. The capacity for compensatory growth is a crop-specific characteristic. The extent to which this capacity is fulfilled depends, among other things, on the spatial pattern of the pest attack. It appears that compensation is less effective if killed or injured plants have an aggregated pattern (Bardner and Fletcher 1974).

Although a great deal of attention has been devoted to the study of spatial pattern in insect populations (Taylor 1984), spatial processes are rarely included in the crop loss assessment models used in crop protection decision-making. This paper describes an attempt to characterize crop responses to patchy pest attack.

Wheat bulb fly (Delia coarctata)

Wheat bulb fly is a stand-reducing pest of wheat (*Triticum aestivum*) crops. Attack by wheat bulb fly larvae on crops at the seedling stage results in aggregations of injured plants, so that the crop consists of a dense stand interspersed with patches where most of the plants have died (Bardner 1968). Attempts to simulate attack on wheat crops by wheat bulb fly larvae in field experiments (Bardner and Huston 1968, Bardner and Fletcher 1969) suggested that compensation for a given level of injury was reduced progressively as the pattern of reduction in density became increasingly aggregated.

McKinlay (1981) carried out a series of field experiments from which he was able to characterize a relationship between wheat yield and wheat bulb fly attack. In each of 49 plots, not treated with insecticide, $r = 6$ independent samples of 30cm of row were scored for the total number of shoots (n_i) and the number of shoots injured by wheat bulb fly attack (z_i). Pest incidence (the proportion of attacked plants in a sample) is $q_i = z_i/n_i$ and average pest incidence in each plot is given by $q = \sum_{i=1}^{r} z_i / \sum_{i=1}^{r} n_i$. At the end of the growing season, an overall estimate of the yield per unit area (y) from each plot was recorded. A simple statistical model of the form $y = cp$, where $p = 1 - q$ and is the average proportion of uninjured plants and c is a proportionality constant estimated by regression analysis, served to describe the effect of wheat bulb fly attack on wheat yield.

Spatial pattern

Data such as those recorded by McKinlay (1981) contain some information relating to spatial pattern. Snedecor and Cochran (1980) give details of a test to examine whether the true values of the binomial variate q_i vary from sample to sample. This involves calculating:

$$\chi^2_{r-1} = \frac{\sum n_i(q_i - q)^2}{q(1 - q)} \tag{1}$$

χ^2 is essentially a comparison of the observed variance among the q_i with the variance that the q_i would have if they were independent samples from the same binomial distribution. A high value of χ^2 indicates that the true proportions differ from sample to sample (Snedecor and Cochran 1980).

The test of homogeneity of the binomial distribution could be caried out for each individual plot, but it would be more informative to obtain an indication of any overall trend in χ^2 with mean incidence. This may be achieved by plotting a graph of $O = \sum n_i(q_i - q)^2$ against $E = q(1 - q)$. The relationship $O = aE^b$ provides a basis for an empirical description of such data. The coefficients a and b can be estimated from the regression of $\log_{10}(O)$ on $\log_{10}(E)$. The slope of this relationship is an estimate of χ^2, which can thus be written in terms of q:

$$\chi^2 = abE^{b-1} = abq^{b-1}(1 - q)^{b-1} \tag{2}$$

Yield loss

The effect of patchy pest injury of the type inflicted by wheat bulb fly on yield loss (l) is to increase the rate of yield loss per unit injury, compared to homogeneous pest injury. Taking χ^2 as an index of patchiness, the simplest formulation of this is:

$$\frac{\mathrm{d}l}{\mathrm{d}q} = c\chi^2 \tag{3a}$$

in which c is a proportionality constant. Substituting equation 2 into equation 3a gives:

$$\frac{\mathrm{d}l}{\mathrm{d}q} = cabq^{b-1}(1-q)^{b-1} \tag{3b}$$

On integration, this provides a relationship (often referred to as a damage curve) between yield loss and pest incidence of the form:

$$l = cab \int_0^q q^{b-1}(1-q)^{b-1}\mathrm{d}q = cabI_q(b,b) \tag{4}$$

where $I_q(b,b)$ is the incomplete beta-function tabulated by Pearson (1968).

It is interesting to compare the characteristics of the model described by equation 4 with those of a typical regression model such as that described by McKinlay (1981). Whereas the regression approach fixes the form of the damage curve by the choice of statistical model, the damage curve described by equation 4 may be linear or sigmoid, depending on the value of b. The sigmoid form, which arises when $b > 1$, is the general form discussed by Pedigo, Hutchins and Higley (1986), while the linear form may be viewed as a special case which arises when $b = 1$. A preliminary reexamination of the data on which McKinlay's (1981) linear regression was based has given no reason to doubt that $b = 1$ in this case, and thus that the linear model is adequate, but the number of samples per plot ($r = 6$) is rather too small to provide a rigorous test of this.

The regression approach to damage curve characterization requires crop yield data as the independent variable. Equation 4 does not require yield data. Instead, the requirements are an estimate of b, which may often be made from the same data already collected to characterize pest incidence, and an assumption about the dependence of the rate of yield loss on the spatial pattern of pest incidence. Equation 3a embodies the simplest realistic assumption, but the same basic framework can encompass more complex formulations (Hughes 1990).

Discussion

Equation 4 represents a damage curve which describes a yield response to the level and spatial pattern of pest incidence. Alternatively, crop loss assessment models may describe a yield response to pest density (the number of pest organisms in a sample). Regression analysis of yield on pest density is a frequently adopted approach, and has been used by Raw and Lofty (1957) and Bardner, Maskell and Ross (1970) to characterize linear relationships between wheat yield and density of wheat bulb fly larvae. Once again this approach ignores spatial processes.

123

Crop responses to pest density include two distinct spatial processes: the effects of patchiness on intraspecific competition in the crop and in the pest population. As discussed above, patchy pest attack may reduce compensatory growth made by surviving healthy plants in a crop. But a greater proportion of a crop may escape the effects of a more patchy pest attack. Hughes and McKinlay (1988) developed a conceptual framework for the integration of these two processes in crop loss assessment models. It requires a model of the yield response to pest incidence such as that developed here, and a relationship between pest incidence and pest density.

If a pest attacks a crop at random, then incidence at mean pest density m is given by $1 - e^{-m}$, where e^{-m} is the proportion of the crop excaping injury as given by the zero term of the Poisson distribution (Justesen and Tammes 1960). Bardner and Lofty (1971) suggest that wheat bulb fly larvae can be regarded as random in their distribution, in which case the incidence-density relationship based on the Poisson distribution seems to be the appropriate choice. More generally, however, insect pest populations may have aggregated spatial patterns, and the degree of aggregation may vary with mean density.

One approach to this problem would be to look for an appropriate aggregated distribution to describe pest incidence. Alternatively, it may be worth investigating the possibility of describing incidence-density curves in terms of a relationship equivalent to that of equation 3a for yield loss-incidence curves. The simplest form would be:

$$\frac{dq}{dm} = cD^{-1} \tag{5a}$$

where D is an index of spatial pattern which increases with increasing aggregation and c is a proportionality constant. For example, taking the variance-mean ratio s^2/m as the index of aggregation:

$$\frac{dq}{dm} = c\left(\frac{s^2}{m}\right)^{-1}. \tag{5b}$$

Taylor's (1961) power law, $s^2 = am^b$ where a and b are regression coefficients, suggests that:

$$\frac{dq}{dm} = ca^{-1}m^{1-b} \tag{5c}$$

in which case:

$$q = \frac{cm^{2-b}}{a(2-b)} \tag{6}$$

assuming that $q = 0$ when $m = 0$.

When $b = 1$, as for wheat bulb fly larvae (Bardner and Lofty 1971), equation 6 implies that mean incidence is directly proportional to mean density. This would have to be the case for the relationships between yield loss and pest incidence (McKinlay 1981) and yield loss and pest density (Raw and Lofty 1957, Bardner, Maskell and Ross 1970) both to be linear, as suggested by the experimental studies. However, since the form of equation 6 is only realistic when $b < 2$, it is clearly not a general model and further investigation of equation 5 is required. In this context, the use of other indexes of aggregation may be of interest. Lloyd's (1967) index of mean crowding can be written in terms of mean density (Iwao 1968), as can the negative binomial index k^{-1}, either from its relationship to Taylor's (1961) power law (Taylor, Woiwod and Perry 1979) or empirically (for example, Logan 1981), so both are candidates for further study.

The conclusion of the present study is that it is possible to formulate qualitative models of crop responses to patchy pest attack using an index approach to characterize spatial pattern. The next step is to develop quantitative versions applicable to particular pest-crop interactions.

Acknowledgments. This paper was presented to the workshop on Spatial Processes in Plant Communities held at Liblice, 18–23 September 1989. The author's visit to Czechoslovakia was financed by the Cultural Exchange Programme of the British Council.

References

Bardner R. (1968): Wheat bulb fly, Leptohylemyia coarctata Fall., and its effect on the growth and yield of wheat.—Annals Appl. Biol. 61: 1–11.

Bardner R. and Flechter K.E. (1969): The distribution of attacked plants.—In: Rothamsted Experimental Station, Report for 1969, Part 1, pp. 199–200. Harpenden: Lawes Agricultural Trust.

Bardner R. and Fletcher K.E. (1974): Insect infestations and their effect on the growth and yield of field crops: a review.—Bull. Entomological Research 64: 141–160.

Bardner R. and Huston P.J. (1968): Effect of gaps on yield.—In: Rothamsted Experimental Station, Report for 1967, pp. 207–208. Harpenden: Lawes Agricultural Trust.

Bardner R. and Lofty J.R. (1971): The distribution of eggs, larvae and plants within crops attacked by wheat bulb fly Leptohylemyia coarctata (Fall.).—J. Applied Ecology 8: 683–686.

Bardner R., Maskell F.E. and Ross G.J.S. (1970): Measurements of infestations of wheat bulb fly, Leptohylemyia coarctata (Fall.), and their relationship with yield.—Plant Pathology 19: 82–87.

Hughes G. (1990): Characterizing crop responses to patchy pathogen injury.—Plant Pathology 39: 2–4.

Hughes G. and McKinlay R.G. (1988): Spatial heterogeneity in yield-pest relationships for crop loss assessment.—Ecological Modelling 41: 67–73.

Iwao S. (1968): A new regression method for analyzing the aggregation pattern of animal populations.—Researches on Population Ecology 10: 1–20.

Justesen S.H. and Tammes P.M.L. (1960): Studies of yield losses I. The self-limiting effect of injurious of competitive organisms on crop yield.—Tijdschrift voor Plantenziekten 66: 281–287.

Lloyd M. (1967): 'Mean crowding'.—J. Animal Ecol. 36: 1–30.

Logan P.A. (1981): Estimating and projecting colorado potato beetle density and potato yield loss.—In: Lashomb J.H. and Casagrande R. [eds.], Advances in Potato Pest Management, pp. 105–118. Stroudsburg: Hutchinson Ross.

McKinlay R.G. (1981): Insecticidal toxicity and cereal grain yield.—Annals Applied Biology 97: 253–256.

Pearson K. (1968): Tables of the Incomplete Beta-Function.—(2nd edition). Cambridge: Cambridge University Press.

Pedigo L.P., Hutchins S.H. and Higley L.G. (1986): Economic injury levels in theory and practice.—Annual Review of Entomology 31: 341–368.

Raw F. and Lofty J.R. (1957). Estimating crop losses due to wheat bulb fly.—Plant Pathology 6: 51–56.

Snedecor G.W. and Cochran W.G. (1980): Statistical Methods.—(7th edition). Ames: Iowa State University Press.

Taylor L.R. (1961): Aggregation, variance and the mean.—Nature 189: 732–735.

Taylor L.R. (1984): Assessing and interpreting the spatial patterns of insect populations.—Annual Review of Entomology 29: 321–357.

Taylor L.R., Woiwod I.P. and Perry J.N. (1979): The negative binomial as a dynamic ecological model for aggregation, and the density dependence of k.—J. of Animal Ecol. 48: 289–304.

Willey R.W. and Heath S.B. (1969): The quantitative relationships between plant population and crop yield.—Advances in Agronomy 21: 281–321.

Krahulec F., Agnew A.D.Q., Agnew S. & Willems J.H. [eds.]: *Spatial processes in plant communities.* pp. 127–143.

Competition process and spatial pattern formation in a Betula ermanii population

Toshihiko Hara[1,2], Hans van Rijnberk[1,3], Heinjo During[1], Masayuki Yokozawa[4] and Kihachiro Kikuzawa[5]

1) *Department of Plant Ecology, University of Utrecht, Lange Nieuwstraat 106, 3512 PN Utrecht, The Netherlands*
2) Present address: *Department of Biology, Faculty of Science, Tokyo Metropolitan University, Tokyo 158, Japan*
3) Present address: *Department of Evolutionary Biology, University of Leiden, Schelpenkade 14A, 2313 ZT Leiden, The Netherlands*
4) *National Institute of Agro-environmental Sciences, Kannondai 3-1-1, Tsukuba 305, Japan*
5) *Hokkaido Forest Experiment Station, Bibai, Hokkaido 079-01, Japan*

Keywords: *Betula ermanii* Cham., Early successional species, Intraspecific competition, Growth dynamics, Size structure, Size variability, Spatial pattern formation, Moran's r.

Abstract. The regeneration process of a natural population of birch (*Betula ermanii*), an early successional species, was investigated in terms of the growth patterns of individual trees and the spatial pattern formation. These two kinds of analyses revealed that size variability is less accentuated in tree height distribution than in DBH (diameter at breast height) and trunk volume, and that individual trees of birch give top priority to tree height growth rather than to DBH growth. This allocation pattern of birch is a crucial characteristic of early successional species, which rapidly invade open sites but then are subject to severe competition during the first phase of vegetation development.

Introduction

There have been only a few studies on the processes of growth dynamics during the course of stand development of naturally regenerated tree populations, although many studies have been made with herbaceous plants under experimental conditions (Kikuzawa 1988). In this paper, we analyze processes of regeneration of a natural birch population (*Betula ermanii* Cham.) with special reference to the growth dynamics of individual trees and spatial pattern formation. One of the authors has published the first paper (Kikuzawa 1988) of this serial research work, and the present paper is the second.

In Hokkaido, Japan, birch forms pure, even-aged stands in open sites created by fire or landslides, and in the subalpine zone it occurs as an early successional species (Kikuzawa

127

1987, 1988). Growth of each individual young birch in a young stand was monitored since 1981. The main objective of our study is to investigate the processes of stand development and intraspecific competition during the natural regeneration of birch.

In the previous paper (Kikuzawa 1988), processes of intraspecific competition were analyzed based mainly on the increase of DBH (diameter at breast height) of each tree and the development of size structure of DBH. In this paper, we analyze the growth patterns of trunk volume, DBH and tree height of each individual as affected by intraspecific competition (or interactions between individuals) and the dynamics of stand structures. We employ the diffusion model for analyzing the growth patterns of individual trees (Hara 1984a, b, 1986a, b) and the autocorrelation in space and time presented here as spatial and spatio-temporal correlograms for analyzing the spatial pattern formation.

Study site

Details of the study site and methods are given in the first paper (Kikuzawa 1988). Here we give only the outline. The study site is the Tobetsu Area of Hokkaido State Forest, Japan (lat. 43°30′ N, long. 141°40′ E). The mean annual temperature is 3.5°C (minimum, −13°C; maximum, 18°C) and the annual precipitation is about 1500mm. An open site was created by scarification in 1971 on a nearly flat mountain slope at 400m altitude. Natural invasion of birch seedlings began in 1971 and continued over a five-year period. By 1981 a dense and nearly pure stand of birch saplings, 2–5 m in height, had developed. A few individuals of *Betula maximowicziana* and *Phellodendron amurensis* were found in the stand.

Methods

A 10 × 10 m permanent plot was set up in 1981. The plot was divided into 100 subplots, each 1 × 1 m. Growth of all individual trees above 1.3 m in height in the plot has been recorded and is still being recorded: DBH since 1981, tree height since 1983 and height at the lowest live branch since 1984. Locations of all individuals were also recorded and each individual was assigned the subplot number (from 1 to 100). To estimate trunk volume, we employed allometric relationships between trunk volume, DBH and tree height which were obtained from trees cut near the permanent plot (five trees in 1981 for stem analysis and 197 trees in 1984). The obtained allometric relationships are as follows: for 1981 and 1982, $V = 0.000203(\text{DBH})^{2.26}$, $r = 0.998$; for 1983-1987, $V = 0.2004((\text{DBH})^2\text{H})^{0.8063}$, $r = 0.978$, where V is trunk volume (m^3), DBH is diameter at breast height (m), H is tree height (m), and r is correlation coefficient.

Analysis

Diffusion model

Let $f(t,x)$ be the size distribution function of size x at time t. The dynamics of $f(t,x)$ is described by the diffusion equation (Hara 1984a, b):

$$\frac{\partial f(t,x)}{\partial t} = \frac{1}{2}\frac{\partial^2[D(t,x)f(t,x)]}{\partial x^2} - \frac{\partial[G(t,x)f(t,x)]}{\partial x} - M(t,x)f(t,x) \tag{1}$$

where $G(t,x)$ is the mean growth rate (i.e. mean of size increments per unit time) of individuals of size x at time t, $D(t,x)$ is the variance of growth rates of individuals of size x at time t, $M(t,x)$ is the mortality rate of individuals of size x at time t. If we know the function forms of $G(t,x)$, $D(t,x)$ and $M(t,x)$ together with the boundary conditions of $f(t,x)$ (generally, $f(A,t) = f(B,t) = 0$ for any t where $A \leq x \leq B$) of a plant population, we can predict the dynamics of size structure theoretically, Furthermore, we can investigate the characteristics of species by comparing these functions between species (Hara 1986a,b). Equation (1) describes the dynamics of only one kind of size measure, for example, only plant weight, only stem diameter, or only plant height.

The extended version of equation (1) which incorporates two kinds of size measure at the same time is

$$\frac{\partial f}{\partial t} = \frac{1}{2}\left(\frac{\partial^2[D_{xx}f]}{\partial x^2} + \frac{2\partial^2[D_{xy}f]}{\partial x \partial y} + \frac{\partial^2[D_{yy}f]}{\partial y^2}\right) - \left(\frac{\partial[G_xf]}{\partial x} + \frac{\partial[G_yf]}{\partial y}\right) - Mf \quad (2)$$

where $f = f(t,x,y)$ is the size distribution function of individuals of size x (say, stem diameter) and y (say, plant height) at time t; $D_{xx} = D_{xx}(t,x,y)$ is the variance of growth rates (variance of increments per unit time) of size x of individuals of size x and y at time t; $D_{yy} = D_{yy}(t,x,y)$ is the variance of growth rates of size y of individuals of size x and y at time t; $D_{xy} = D_{xy}(t,x,y)$ is the covariance of growth rates of x and y of individuals of size x and y at time t; $G_x = G_x(t,x,y)$ is the mean growth rate of size x of individuals of size x and y at time t; $G_y = G_y(t,x,y)$ is the mean growth rate of size y of individuals of size x and y at time t; $M = M(t,x,y)$ is the mortality rate of individuals of size x and y at time t. In this paper, x and y represent DBH and tree height respectively. Therefore, $f(t,x,y)dxdy$ is the number of individuals of DBH $\in [x, x+dx]$ and tree height $\in [y, y+dy]$.

Autocorrelation in space and time

To reveal the spatial structuring of tree attributes, the autocorrelogram analysis procedure was performed according to Upton and Fingleton (1985). It calculates autocorrelation coefficients for a number of distance classes. The autocorrelation is the correlation between all possible pairs of cell (= subplot, in our case) values of an attribute at a specified distance. In our case distances ranged from 1 to 5 cells in vertical and horizontal direction (rooks' case). Every cell value thus serves as a 'X' value as well as an 'Y' value just as in the Pearson correlation. The value thus obtained is corrected to minimize calculation biases arising from edge effects in grid recordings (the number of neighbouring cells is limited at the edges). Positive autocorrelation indicates clustering, negative autocorrelation dispersion. A typical correlogram resulting from a checkerboard pattern, the fields of which are for instance 2×2 cells, shows a jaw-teeth scope line with highly positive autocorrelation at distances of 1, 4, 8, 12, ... cells and with highly negative autocorrelation at distances of 2, 6, 10, 14,... cells. The typical cluster size can be found from the number of cells distance with locally extreme negative autocorrelation if it is preceded by positive autocorrelation for smaller distances (2, 6, 10, 14,...). Typical dispersion distances are obtained from the distance with locally extreme positive autocorrelation if it is preceded by negative auto-correlation for smaller distances (1, 4, 8, 12,...). Autocorrelation is calculated as Moran's

129

I,

$$I_d = \frac{N \sum_i \sum_j W d_{ij}(x_i - \overline{x})(x_j - \overline{x})}{S_o \sum_i (x_i - \overline{x})^2} \tag{3}$$

where N is the number of cells ($100(= 10 \times 10)$ subplots, in our case), $W d_{ij}$ is the matrix defining which pairs of cells should be compared and included in the summation. S_o is the S-statistic according to rook's definition of contiguity of the grid cells, x_i is the observed value of an attribute in cell i, x_j is the observed value of an attribute in cell j, $\overline{x}$ is the mean value of attribute x; x_i and x_j are analogous to two variables of the normal Pearson correlation.

Upton and Fingleton (1985) supply formulae for the expected values of I and the standard deviation, from which a standard normal deviate ($[O(I) - E(I)]/sd(I)$) is calculated as Z-value. For random pattern the Z-value follows the standard normal distribution, provided that the number of nonzero grid observations is not too small (see Upton and Fingleton 1985). If the absolute value of Z exceeds 1.96, the 5% significance level, we consider it as significant autocorrelation.

So far the analysis has provided us with the spatial pattern for each separate year. However, by applying the same calculation procedure we can analyze a spatial pattern (distance lag in space) combined with time lag as a spatio-temporal correlogram. Now for every combination of a certain spatial lag (number of cells apart) with a certain temporal lag (number of years apart), autocorrelation is calculated. If the spatial lag (temporal lag) is zero, the autocorrelation gives the temporal pattern within the cell (spatial pattern within the year). The actual calculation is made by varying x_i and x_j as values of different grid cells for the specified time lag (number of years difference). The subscripts i and j now designate a certain cell in a certain year. This is visualised in Figure 1, which shows an example of the combination of 2-cell distance (spatial lag) and 1-year time difference (temporal lag). For the expected value and the standard deviation of I, Upton and Fingleton (1985) gives modified formulae resulting in the new test variable Z (1.96 as 5% significance level).

Because of a bias toward negative spatial autocorrelation over time, which results from a consistent trend in the values of an attribute over time, we decided to adopt the modification as used in van der Hoeven et al. (1990). Instead of using the overall mean, the mean within one year was used in the calculations:

$$I_d = \frac{N \sum_i \sum_j W d, t_{ij}(x_i - \overline{x})(y_j - \overline{y})}{S_o \sum_i (x_i - \overline{x})(y_i - \overline{y})} \tag{4}$$

where N is the number of cells ($100 = 10 \times 10$) subplots in our case), S_0 is the S-statistic according to rook's definition of contiguity of the grid cells, x_i is the observed value of an attribute in cell i in one year, y_j is the observed value of the attribute in cell j in another year, $\overline{x}$ is the mean value of the attribute in the year of value x, $\overline{y}$ is the mean value of the attribute in the year of value y. Every cell value is thus related to the deviation from the mean within its own year.

Because Moran's I can be larger than 1 or smaller than -1, we corrected it as advocated by Upton and Fingleton (1985) to obtain the modified autocorrelation coefficient r

130

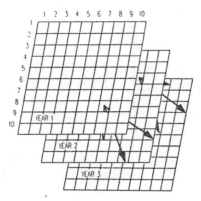

Figure 1. Schematic representation of grids in different years to demonstrate comparison between different cells in different years. Six of the eight comparisons which contribute to calculation of spatio-temporal autocorrelation for 2-cell distance (spatial lag) and 1-year time distance (time lag) are shown for cell (8,8).

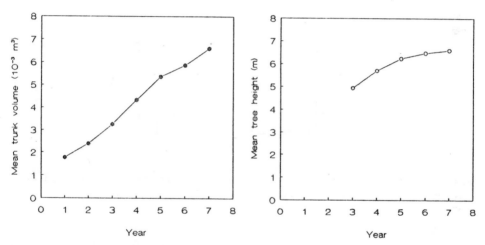

Figure 2. Changes in mean values of trunk volume (left) and tree height (right) of *Betula ermanii* over the period of the present study, 1981 (year 1)–1987 (year 7). The number of trees per plot (100 m^2) was 267 in year 1, 211 in year 3 and 128 in year 7.

which ranges from -1 to $+1$. Autocorrelation in correlograms for each separate year is denoted by r, that in spatio-temporal correlograms by r''.

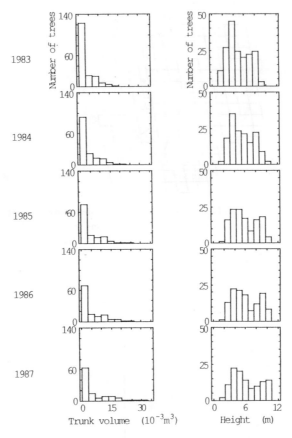

Figure 3. Size frequency histograms of trunk volume (left) and tree height (right) of *Betula ermanii* from 1983 (year 3) through 1987 (year 7).

Results

Size structure

Mean trunk volume and mean tree height increased from year 1 (1981, tree age ca. 10 years) through year 7 (1987, tree age ca. 16 years)(Fig. 2). Figure 3 shows histograms of size frequency distribution of trunk volume and tree height. Those of DBH are given in the previous paper (Kikuzawa 1988). Figure 4 shows changes in CV (coefficient of variation) and skewness of trunk volume and tree height. Trunk volume has much greater size variability and more positively skewed size distribution (L-shaped distribution) than tree height. Tree height has almost symmetrical size distribution and we can clearly see size bimodality in tree height. However, this is not the case in trunk volume (Fig. 3). The

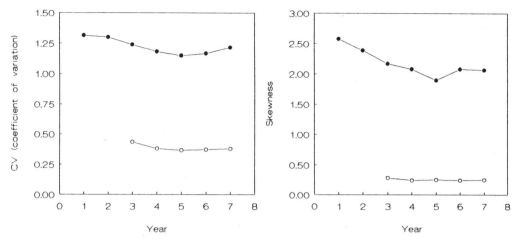

Figure 4. Changes in CV (coefficient of variation)(left) and skewness (right) of trunk volume (closed circle) and tree height (open circle) of *Betula ermanii* from 1981 (year 1) through 1987 (year 7).

previous paper reported size bimodality also in DBH distributions, but size bimodality in tree height is more accentuated than that in DBH.

The previous paper (Kikuzawa 1988) also reported that the $-3/2$ law of self-thinning (Yoda *et al.* 1963) holds in the birch population of the present study, namely $V = 9094N^{-3/2}$ where V is mean trunk volume (m^3) and N is tree density per hectare.

Figure 5 shows the $G(t, x)$ functions of trunk volume and tree height. Figure 6 shows $G_x(t, x, y)$ and $G_y(t, x, y)$ functions where x and y represent DBH and tree height respectively. We clearly see that small-sized suppressed individuals grow only in tree height but not in trunk volume and DBH.

Figure 7 shows the relationship between DBH and tree height and suggests size-dependent D/H ratio (DBH/tree height ratio) which represents slenderness of the tree. Figure 8 shows mortality of each class of D/H ratio. In accordance with size bimodality in tree height distributions (Fig. 3) and the G functions (Figs. 5 and 6), D/H ratio in year 7 also seems to fall into two groups, dominant and suppressed individuals; individuals shorter than ca. 7m have rather constant but low values of D/H ratio, while individuals taller than ca. 7m show an increase in D/H ratio with tree height. However, the overall trend is as follows: large dominant trees tend to have greater D/H ratio and lower mortality; small-sized and slenderer (smaller D/H ratio) trees are subject to higher mortality.

Spatio-temporal pattern formation

Figure 9 shows contour maps of mean trunk volume and mean tree height per subplot of the study plot. The contour maps of trunk volume show steeper slopes (i.e. denser lines) than those of tree height. This means that the difference between maximum and minimum values of mean trunk volume per subplot is greater than that of mean tree height

133

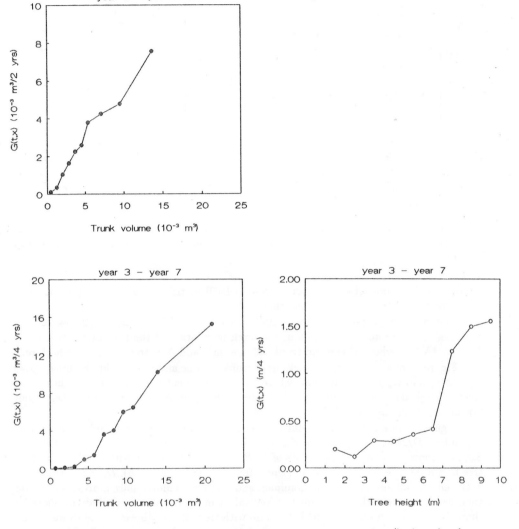

Figure 5. $G(t, x)$ functions (mean growth rate of individuals of size x at time t) of trunk volume (left) and tree height (right) of *Betula ermanii* for 1983–1987 (year 3–year 7). x-axes are sizes at year 3.

per subplot, and mean trunk volume per subplot is more variable from subplot to subplot than mean tree height per subplot. Mean tree height per subplot is rather uniform between subplots.

134

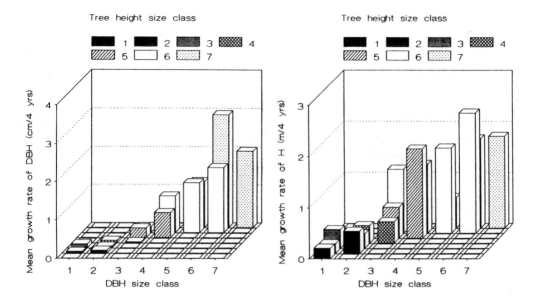

Figure 6. $G_x(t, x, y)$ function (left) and $G_y(t, x, y)$ function (right) for 1983–1987 (year 3–year 7), where x and y represent DBH (diameter at breast height) and tree height of *Betula ermanii* at year 3 respectively. The $G_x(t, x, y)$ function ($G_y(t, x, y)$ function) represents the mean growth rate of size x (size y) of individuals of size x and y at time t. The value of each size class is as follows (from size class 1 to 7): for DBH, 1.09, 2.28, 3.46, 4.65, 5.84, 7.02, 8,21 cm; for tree height, 1.64, 2.93, 4.21, 5.50, 6.79, 8.07, 9.36 m.

The results of the analysis of pattern in values of mean number of trees, mean trunk volume and mean tree height per subplot are shown in the spatial and spatio-temporal correlogram of Figures 10 and 11. These results, especially spatio-temporal correlograms (Fig. 11), support the qualitative observation of the contour maps in a statistical manner. The spatial correlograms (Fig. 10) display spatial patterns within the set of observations of each year separately. The patterns of trunk volume and height are rather similar, with low (although not significant) values for adjacent cells (spatial lag 1, $d = 1$), highest and often significant values at $d = 3$ (spatial lag 3), and low values again at $d = 4$ (significant in the case of tree height). This suggests clustering with clusters of about four by four units (cells). In the course of time, the pattern tends to become more pronounced. Thus, in the cases of trunk volume and tree height the spatial correlograms (Fig. 10) clearly indicate some pattern, but only a few of the r values are significant. This is partly due to the fairly low sample size.

The r'' values of the spatio-temporal correlograms (Fig. 11) at time lag zero can be seen as averages of the spatial correlograms (Fig. 10) over the years of observation. In the cases of trunk volume and tree height, these values show highly significant clustering with cluster size $d = 4$, which reinforces the interpretation of the spatial correlograms of the individual years (Fig. 10). Moreover, the trend toward low values at $d = 1$ observed

135

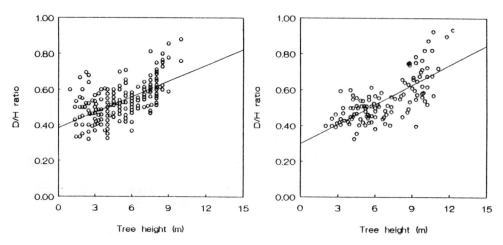

Figure 7. Relationships between DBH (diameter at breast height) and tree height on a log-plot in 1983 (year 3)(left) and 1987 (year 7)(right) of *Betula ermanii*. The regression curve is given by the generalized allometric equation with $h = 1$, $1/H = 1/(AD^h) + 1/H_{max}$, where H is tree height. D is DBH, A and H_{max} are constants specific to the stand age (e.g. Ogawa and Kira 1977). In 1983, $A = 2.623$, $H_{max} = 18.213$, $n = 211$; in 1987, $A = 2.779$, $H_{max} = 21.171$, $n = 128$.

in the spatial correlograms (Fig. 10) is now significant only in the case of the spatio-temporal correlogram of trunk volume (Fig. 11), which indicates some regularity in the distribution at this scale. This reflects the trend that cells with trees of large trunk volume are surrounded by cells with low values of this measure. This trend is, however, not clear in tree height (nonsignificant at $d = 1$, although r'' values are negative).

The spatio-temporal correlograms also show high autocorrelation at increasing time lags, indicatging a strong carry-over effect of the pattern through time; the largest trees remain the largest, and in cells with small trees these will remain relatively small.

In striking contrast to these patterns is the pattern of the number of individuals per cell (= subplot). The spatial correlogram (Fig. 10) indicates a tendency toward clustering at $d = 1$ and to regularity at $d = 2$ and $d = 5$ in the early years, but in the later years these patterns disappear completely; the numbers of trees per cell are randomly distributed at all scales analyzed. The spatio-temporal correlogram (Fig. 11) similarly suggests some regularity in distribution at scales $d = 2$–3 and $d = 5$, but the pattern did not develop strongly, and the carry-over of the pattern through time is much less than in the cases of trunk volume and tree height.

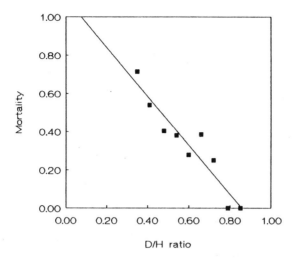

Figure 8. Relationship between mortality and D/H ratio (DBH/treeheightratio) of *Betula ermanii*. Mortality is calculated between 1983 (year 3) and 1987 (year 7) for each D/H ratio class. (mortality) $= 1.097 - 1.282(D/H)$, $n = 9$, $r = -0.946$.

Discussion

In this paper, we concentrate on only the $G(t,x)$, $Gx(t,x,y)$ and $Gy(t,x,y)$ functions which represent basic growth characteristics of the species (Hara 1984a,b; West, Jackett and Borough 1989), although the $D(t,x)$ functions play an important role in some cases, especially in the early growing stages or widely-spaced populations (Hara 1984a,b; Petersen 1988).

The previous paper (Kikuzawa 1988) reports that the plot of the present study contains at least three cohorts of different age classes. This is clear from the marked bimodality in tree height distributions (Fig. 3) and the crown depth diagrams in the previous paper (Fig. 4 in Kikuzawa 1988). However, size bimodality is not clear in trunk volume distributions. We see a clear difference in the individual growth pattern between tree height and trunk volume (or DBH) (Figs. 5 and 6). In accordance with bimodality in tree height distribution, we can recognize two groups of individual trees in the $G_x(t,x,y)$ and $G_y(t,x,y)$ functions in Figure 6, i.e. small suppressed trees and large dominant trees. In the plot of the present study, small suppressed individuals (left mode in the tree height distribution) belong to the young cohort, and large dominant ones (right mode in the tree height distribution) to the old cohort. Small suppressed individuals grow only in height but not in DBH. Large dominant individuals grow not only in height but also in DBH, in fact almost all trees, small and large, grow in height but only large ones grow in DBH. These growth patterns of DBH and tree height lead to the growth pattern of trunk volume; small suppressed individuals grow disproportionately little as compared with large dominant ones (Fig. 5), because trunk volume is approximately proportional to $(DBH)^2 \times$ (treeheight)(see the Methods).

137

Figure 9. Contour maps of mean trunk volume per subplot (left) and mean tree height per subplot (right) of *Betula ermanii* for several observation years. Coordinates (1,1), (2,1), ... (10,10) in the contour maps correspond to subplot 1,2, ... 100, respectively.

138

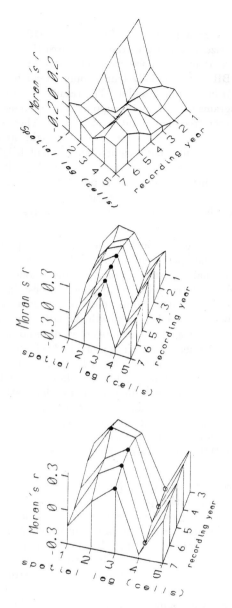

Figure 10. Spatial correlograms based on Moran's r of number per subplc 1–year 7)(top), mean trunk volume per subplot for 1981–1987 (yea and mean tree height per subplot for 1983–1987 (year 3–year 7) (l *manii*. Closed and open circles represent positive and negative corr 5% significant level. Other points are nonsignificant.

139

The differences in the growth patterns of tree height, DBH and trunk volume brings about the difference in size variability of these measures (Hara 1984a,b, 1986a,b); tree height tends to have less accentuated size variability (size hierarchy or "dominance and suppression"), while DBH and especially trunk volume tend to have more accentuated size variability as shown in Figures 3 and 4. The results of contour maps (Fig. 9) and spatio-temporal correlograms (Fig. 11) also support this difference in the growth pattern and size variability between tree height and trunk volume. Therefore, effects of one-sided competition (asymmetric competition or "competition for light")(e.g. Weiner and Thomas 1986) would be much clearer in plant weight than in plant height. In this sense, size bimodality itself does not always mean greater size variability, although it would be true that bimodality shows two distinct classes in the population, i.e. dominant and suppressed individuals (e.g. Ford 1975). In our case, dominant individuals and suppressed ones are regarded as belonging to the old cohort and the young one respectively (e.g. Huston and DeAngelis 1987).

From the above results, we can conclude that the growth pattern of individual birch trees as an early successional species under crowded conditions is as follows (Fig. 12). Tree height of each individual in a stand determines the amount of light each individual receives, and therefore dry matter production by photosynthesis in the canopy. Dry matter production means growth in weight (or trunk volume). The next step is the partitioning of dry matter production between tree height growth and DBH growth during the vegetative growth, which is specific to the species (at present, we have no data about the allocation to root growth). Our results show that birch individuals give top priority to tree height growth. Large individuals, which gain a relatively large amount of dry matter production because they are tall in the stand, can afford to grow in both tree height and DBH. Small individuals, which gain only a relatively small amount of dry matter production especially in planophile plants because they are short in the stand (Kuroiwa 1960; Hara 1986a), can grow only in tree height because they cannot afford to grow in both DBH and tree height. Therefore, small individuals tend to be slenderer (smaller D/H ratio) than large ones, and individuals with smaller D/H ratios are much more subject to mortality because of their slenderness.

Whereas the patterns of mean trunk volume and tree height became more pronounced with time, the pattern of tree number per subplot disappeared completely in the later years of observation, presumably as a result of the density-dependent mortality observed in the plot (Kikuzawa 1988). The increase of pattern in trunk volume and height through time, as suggested by the spatial correlograms (Fig. 10), may be in part due to mortality of the smallest individuals; in cells originally containing two or more individuals, the mean size increases when the smallest tree dies, and this effect will be strongest when the survivor is largest. However, effects of differences in growth due to initial size differences and one-sided competition for light will be more important causes of the increasingly pronounced spatial patterns over the years.

The growth pattern in favor of height growth was reported in annual plants under experimental conditions (Hara 1984a, b). In such experimental and strictly even-aged populations, this characteristic of growth pattern gives the straight line $G(t, x)$ function of plant height (Hara 1984a,b). Under natural conditions, however, most populations consist

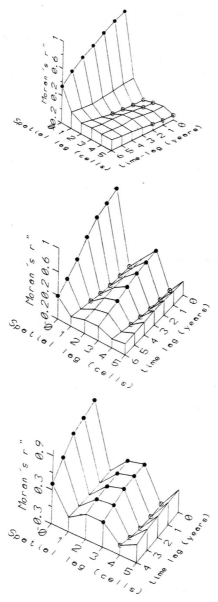

Figure 11. Spatio-temporal correlograms based on Moran's r'' of number per subplot for 1981–1987 (year 1–year 7) (top), mean trunk volume per subplot for 1981–1987 (year 1–year 7) (middle), and mean tree height per subplot for 1983–1987 (year 3–year 7) (bottom) of *Betula ermanii*. Closed and open circles represent positive and negative correlation respectively at 5% significant level. Other points are nonsignificant.

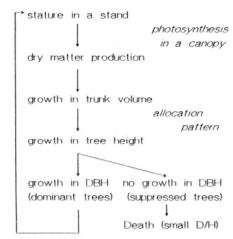

Figure 12. Individual growth pattern of *Betula ermanii*, an early successional species, under crowded conditions.

of several cohorts of different ages and each cohort may have straight line $G(t, x)$ functions of plant height with different slopes as can be seen in Figure 5.

We think that the growth pattern of birch in which each tree gives top priority to height growth rather than DBH growth is an important characteristic of birch as an early successional species, which invades an open site as soon as ones created, because in this way the trees can overtop and outcompete other plant species in terms of canopy photosynthesis. In this sense, this allocation pattern between tree height growth and DBH growth can be regarded as a crucial element in the life history strategy of birch.

We should note, however, that this allocation pattern is advantageous only for large dominant individuals of birch if intraspecific competition is severe. It is rather disadvantageous for small suppressed individuals of birch, because small individuals, which still retain this allocation pattern, are subject to high mortality due to slenderness caused by this same pattern.

Acknowledgements. This study was done while the first author (T.H.) was at the Department of Plant Ecology, University of Utrecht, as a visiting research fellow in 1989, and he thanks all the members of this Department for their hospitality.

References

Ford E.D. (1975): Competition and stand structure in some even-aged plant monocultures.—J. Ecol. 63: 311–333.

Hara T. (1984a): A stochastic model and the moment dynamics of the growth and size distribution in plant populations.—J. Theor. Biol. 109: 173–190.

Hara T. (1984b.): Dynamics of stand structure in plant monocultures.—J. Theor. Biol. 110: 223–139.

Hara T. (1986a): Growth of individuals in plant populations.—Ann. Bot. 57: 55–68.

Hara T. (1986b): Effects of density and extinction coefficient on size variability in plant populations.—Ann. Bot. 57: 885–892.

Huston M.A. and DeAngelis D.L. (1987): Size bimodality in monospecific populations: a critical review of potential mechanisms.—Amer. Nat. 129: 678–707.

Kikuzawa K. (1987): Regeneration and stand-density control for young birch stands in subalpine zone of Hokkaido.—In: Fujimori T. and Kimura M. [eds.], Human impacts and management of mountain forests, pp. 305–314. Forest and Forest Products Research Institute, Ibaraki, Japan, 421 pp.

Kikuzawa K. (1988): Intraspecific competition in a natural stand of *Betula ermanii.*—Ann. Bot. 61: 727–734.

Kuroiwa S. (1960): Intraspecific competition in artifical sunflower communities.—Bot. Mag., Tokyo, 73: 300–309.

Ogawa F. and Kira T. (1977): Methods of estimating forest biomass.—In: Shidei T. and Kira T. [eds.], Primary productivity of Japanese forests, pp. 15–25. University of Tokyo Press, Tokyo, 289 pp.

Petersen T.D. (1988): Effects of interference from *Calamagrostis rubescens* on size distributions in stands of *Pinus ponderosa.*—J. Appl. Ecol. 25: 265–272.

Upton G.J.G. and Fingleton B. (1985): Spatial data analysis by example. Volume 1. Point pattern and quantitative data.—-Wiley, Chichester.

van der Hoeven E.C., de Kroon H. and During H.J.D. (1990): Fine scale spatial distributions of leaves and shoots of two chalk grassland perennials.—Vegetatio (in press).

Weiner J. and Thomas S.C. (1986): Size variability and competition in plant monocultures.—Oikos 47: 211–222.

West P.W., Jackett D.R. and Borough C.J. (1989): Competitive processes in a monoculture of *Pinus radiata* D.Don.—Oecologia 81: 57–61.

Yoda K., Kira T., Ogawa H. and Hozumi K. (1963): Self-thinning in overcrowded pure stands under cultivated and natural conditions.—J. Biol., Osaka City University, 14: 107–129.

Krahulec F., Agnew A.D.Q., Agnew S. & Willems J.H. [eds.]: *Spatial processes in plant communities.* pp. 145–159.

Coexisting competitors in a sedge-grass marsh, Central Europe

Lenka Soukupová

Institute of Botany, Czechoslovak Academy of Sciences, Dukelská 145, CS-379 82 Třeboň, Czechoslovakia

Keywords: *Caricetum gracilis, Carex acuta, Carex vesicaria, Calamagrostis canescens,* Habitat partition, Interspecific competition, Water level, Horizontal spreading, Timing of short life-cycle

Abstract. Seasonal dynamics of the *Caricetum gracilis,* a fen community in the Třeboň Biosphere Reserve, has been analysed with regard to the life history of graminoids, and habitat terrestrialization. Two tall sedges, *Carex gracilis* and *C. vesicaria,* and a loosely tufted *Calamagrostis canescens,* create a skeletal pattern invaded by accessory plant populations. Biomass production in the three-year old monocultures of *Carex gracilis, C. vesicaria* and *Calamagrostis canescens,* was 1.74, 1.18 and 1.42 kg.m^2, respectively, suggesting strong competition within the community. High tolerance to waterlogging enables both *Carex* species to co-exist in a stable equilibrium in habitats exposed to inundations. However, their life histories differ substantially: The less vigorous *C. vesicaria* produces larger number of seeds, creates overwintering shoots, and expands by side-by-side tillers in a 'phalanx way' in the period when 'guerilla-dispersed' shoots of *C. gracilis* are established. Allocation of the aboveground (S) and root biomass (R) prefers *Calamagrostis canescens* (R/S = 1.2) in competition with *Carex gracilis* (R/S = 3.7), enabling the grass to invade thinned tops of sedge hummocks in the course of terrestrialization. However, joint occurrence of the both species results in a non-equilibrium coexistence when the growth of *Calamagrostis canescens* is prevented by higher water level and the sedges become dominant.

Introduction

Microscale pattern in communities of sessile organisms, such as plants, may result from invisible phytosociological interactions among coexisting populations (Kershaw 1973). Braakhekke (1980) distinguished between the stable-equilibrium and non-equilibrium coexistence. The former category denotes a state when competitive exclusion is avoided as a result of different inherent life strategies. The latter category was introduced to describe results of controversial processes which are connected with different sources of variation and disturbance, and therefore their effect on community pattern is usually indirect. Moreover, a concept of the non-coexistence equilibrium was proposed for a certain kind of coexistence when individual species occupy distinct microsites within a heterogeneous environment, so that their interference is only marginal (see also regeneration niche by Grubb 1977).

Five categories of plant traits that enable stable-equilibrium coexistence of populations can be recognized: (a) growth parameters, (b) niche differentiation, (c) interference relationships, (d) timing of life histories, and (e) genetic constitution (inbreeding, hybridization). Few attention has been devoted to the effects of timing in life cycles (Soukupová 1988). Therefore in the following study coexistence is discussed with regard to interlocking of life cycles, growth morphology, and adaptability among plant species in a community.

Material

To reveal the controls of spatial pattern by biotic factors a simple mature community had to be found, in particular one with only slight environmental heterogeneity and consisting of only a few populations. A wetland community was chosen on a 1.0 m deep peat where repeated annual inundations smooth spatial differences in availability of nutrients. The *Caricetum gracilis* Almquist 1929, a sedge-grass fen association widespread in river floodplains and pond/lake littorals of temperate Europe (Blažková 1971), was analysed within the 'Rožmberk project', an integrated ecological research programme, run in the Třeboň Biosphere Reserve, SW Czechoslovakia (Jeník and Květ 1983). The investigated association was marked by a low Shannon-Wiener index of diversity ($1.1 \leq H'(2) \leq 1.8$), intermediate equitability ($0.45 \leq E \leq 0.65$), and a low index of dominance ($0.3 \leq C \leq 0.6$); number of species ranged between 5.5 ± 0.6 and $7.8 \pm 1.3 \mathrm{m}^2$ ($n = 12$), and the aboveground dry biomass varied between 368 and 536 g.m^{-2}. Phytosociological relevés are given by Hroudová (1988).

Along the examined hydrosere the sedge-grass fen association contacts the *Acoretum calami* Eggler 1933 on the wetter side; in more eutrophicated conditions, the *Glycerietum maximae* Hueck 1931 or stands dominated by *Galium elongatum* (Soukupová 1986) are adjacent. Large-scale development of wetlands with *Caricetum gracilis* in the hydrosere has taken place on the shores of the Rožmberk fishpond since its foundation in 1590. Successively tall sedges and reed *Phragmites australis* spread largely over the irregularly mown fen.

Methods

Several procedures were combined while describing and detecting vegetation pattern in the *Caricetum gracilis*. Field observations were complemented with experimental cultivations in a semi-controlled and/or man-induced environment.

Linear transects. — Species frequency was assessed in September 3–6, 1982, along two point linear transects which crossed through the hydrosere of the Rožmberk fishpond. The basic transect was 120 m long and outwards from the shore from the 360 m to the 480 m mark; intersecting this at 450 m mark was the transverse transect which ran parallel to the shore. Readings were taken at 0.05 m intervals where basal parts of plants touched the right margin of a metal scale. Within 1 m long intervals the frequency of species was then counted. On October 20, 1982, and April 24, 1983, samples of ground water were collected from probes bored at 30 m intervals, and their chemistry was analysed according to Hofman (1965). Growth potential was obtained by cultivation on solid media for 1005 hrs. (Lukavský 1983).

Spatial analysis. — Distribution of different stands within the *Caricetum gracilis* association was mapped on July 30, 1982 (Fig. 2). Microscale pattern and its changes

146

were examined on permanent plots 1 × 1m between April 1981 and July 1983. Micromaps were recorded on April 10, 1981, on May 5, 1982, and on May 3, 1983, in the flowering stage of sedges, using a grid with nylon strands of mesh 0.1 × 0.1m in a wooden frame.

Structure analysis. — Inclined point quadrat, a non-destructive method (Warren Wilson 1959), was applied in biomass evaluation of the above mentioned permanent plots. Measurements continued from June 23 to July 5, 1981, from July 10 to July 20, 1982, and from June 21 to June 29, 1983, at the seasonal peak of biomass production in the *Caricetum gracilis*.

Cultivation. — Experiments were carried out to analyse (i) spreading of plants, (ii) the effect of waterlogging, and (iii) interspecific competition. All treatments ran outdoors during the 3-year period from May 15, 1981 to September 2, 1983. Single, 1-year-old vegetative shoots were used as initial offspring; their fresh mass was 5.3 ± 3.1g, 5.3 ± 2.0 g, and 2.1 ± 0.8g (n = 15), in *Carex gracilis, C. vesicaria* and *Calamagrostis canescens*, respectively. The former treatment (i) was carried out in situ, in beds inside natural stands near the Rožmberk fishpond, the latter two treatments (ii+iii) were performed in experimental containers in the grounds of the Institute of Botany, Třeboň, Czechoslovakia.

Field beds near the Rožmberk fishpond were established thus: a trench in the peat was dug out to 1 m depth, and filled inversely by excavated topsoil with the tussocks and roots covered by a layer of the excavated subsoil; a black plastic foil protected beds against root invasion from the adjacent stands; in order to preserve normal fluctuations of the ground water level, the foil was perforated at the bottom. Initial planting density of the shoots was 1 per m².

Experimental tanks (1 × 1 × 0.5m) were filled with sand. During the 1982 season 5l of a nutrient solution (after Dykyjová and Véber 1978) was successively applied to each tank (for details see Soukupová 1986). Water level was manipulated to simulate three different states of wetland habitat: (i) terrestrial — with water level reaching from −0.20 to −0.12m, (ii) limosal — water level between −0.05 and +0.05m, and (iii) littoral — water level between +0.15 to 0.20 m; the habitats are named according to Hejný and Husák (1978). Initial planting density was 24 shoots per tank, and it was intended to gain stands similar to those in field within a year; to test interspecific competition between *Calamagrostis canescens* and *Carex gracilis* a mixture of 12 :12 shoots per tank was studied.

Seed germination. — Seeds collected on 21 May, 1982, were tested for germination in Petri dishes on filter paper, under fluctuating day and night temperature of 22 and 10°C, and 520 W m⁻² light intensity situated in a phytotron KTLK 1250, VEB Nema Netschau. The effect of lower light intensities from 5 to 70 W m⁻², and of constant temperatures ranging from 20 to 40° C, were tested on gradient plates, a device designed by Lukavský (1975).

Statistical data. — The expression $\bar{x} \pm s$ refers to mean ± standard deviation. Principal component analysis (PCA) using Euclidean distance was carried out (Orlóci 1975); HP-85 was used for computations.

147

Results and Discussion

Community pattern and its dynamics

Along the linear transect through the investigated association, *Carex gracilis* was noted as the most frequent species accompanied either by *C. vesicaria*, or by *Calamagrostis canescens* (Fig. 1 — bottom). According to the proportions of these species, three kinds of stand (= facies) were distinguished within the community: facies with strongly dominant *Carex gracilis* (hereafter Car-grac-facies), facies with an equal share of both *Carex* species (hereafter Car-ves-facies) and facies with dominant *Calamagrostis canescens* and subordinate *C. gracilis* (hereafter Cal-can-facies). Distinct boundaries were observed between the Cal-can-facies and Car-grac-facies (Fig. 2). The Car-ves-facies was distributed merely within the Car-grac-facies, its appearance on the transects being evident in segments from 426 to 430 m and from 460 to 463 m on the basic transect across the hydrosere, and from 67 to 70 m and from 94 to 96 m on the transverse transect.

Gradient analysis has not revealed any significant difference in chemical composition of ground water (maximum values are given): pH = 5; $\simeq 60\ \mu$ g.l^{-1} P-PO$_4$; 400 to 700 μ g.l^{-1} N-NH$_4$ (higher spring values in Fig. 1 were caused by recent enrichment); 150 to 200 μ g.l^{-1} N-NO$_3$ (in autumn); 3.3 and 8.6 mg.l^{-1} Na$^+$ in spring and autumn, respectively; 30 to 40 mg.l^{-1} Ca^{++} (in spring); $\simeq 5$ mg.l^{-1} Mg^{++}; 2.0 to 2.5 mg.l^{-1} K$^+$ (in autumn)(Fig. 1 — top right). Also the soil profile was homogeneous beneath the community; 0.00–0.05 m: man-induced surface horizon, 0.05–0.10 m: flood deposits, 0.10–0.25 m: carbonized sedge peat, 0.25–0.80: sedge peat (with wood remnants), 0.80–1.00 m: grey silt, > 1.0 m bedrock sediments.

Annual changes in water level were similar around the examined community (Fig. 1 — top left), except that a shorter inundation period was observed on the drier margin (450 to 480 m) where Cal-can-facies is more important. Inundations appear in spring and reach 0.1 m above soil surface in general. In late summer water level sinks to its lowest value but is usually not more than 0.65 m below ground. While monitoring fluctuations of water level from 1979 to 1983 (Přibáň in prep.) gradual terrestrialization was revealed on the investigated biotope. In 1979, for example, the water level sank no more than 0.15 m below ground level, while in 1983 it reached lower than 0.4 m. Also duration of floods shortened, from several weeks in 1980 to none in 1983. Thus on the shore of Rožmberk fishpond, primarily the hydrological regime seemed to control the development of the *Caricetum gracilis*.

Time analysis indicated changes in vegetation composition on permanent plots from 1981 to 1983. PCA ordination of quadrats from the different facies projected them into three clusters in 1981 (Fig. 3A). In 1983 (Fig. 3B) evaluation of the same quadrats showed a continuous cluster corresponding to the transition of the Car-grac-facies to Cal-can-facies (quadrats of Car-ves-facies still formed a separate cluster). The alteration of the stand dominated by *Carex gracilis* was enhanced after successful establishment of *Calamagrostis canescens* which followed an excessive disturbance when sedge tussocks became tattered during winter inundation by frost and ice in 1982 (Fig. 4: G). As documented by micromaps in Fig. 4: C, the same disturbance caused only annual fluctuations in vegetation composition within the other two facies.

The occurrence of new shoots of *Calamagrostis canescens* on plots in Cal-can-fac was

148

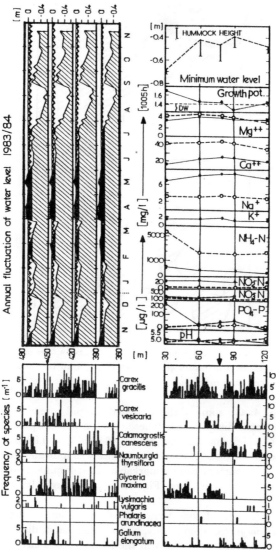

Fig. 1. Spatial heterogeneity in the investigated *Caricetum gracilis* community. Bottom — composition of vegetation expressed as species frequency on two linear transects: each column displays 1-m-long segment (point of intersection is indicated by arrows, for details see Methods section). Left — basic transect outwards from the shore of Rožmberk fishpond (distance is given). Right — transverse transect situated parallel to and 450 m away from the shoreline; the furthest point represents the centre of the community. Top — Left: water level with regard to soil surface on the basic transect (vertical lines — littoral conditions, solid black — limosal conditions, diagonal lines — terrestrial conditions). Top — Right: hydrochemical analysis, growth potential and status of water level on the transverse transect (solid line and closed symbols — October 20, 1982, broken line and open symbols — April 24, 1983).

149

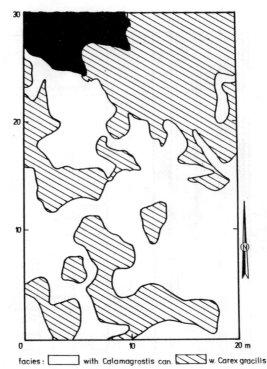

facies: [____] with Calamagrostis can. [\\\\\\] w. Carex gracilis

Fig. 2. Differentiation of the *Caricetum gracilis* community into two facies in 1982 (solid black
— association *Glycerietum maximae*).

only temporary (Fig. 4: C): some of them created a rapid spread of the invading stand,
but the more distant ones disappeared. These changes were displayed also by increased
index of dominance *C* which changed, e.g., from 0.46 in 1981 to 0.69 in 1983 on the quadrat
in Fig. 4: C. Although no convergence or divergence was indicated in the Car-ves-facies,
its changes of microscale pattern were remarkable. The dominant species shifted among
different microsites in successive years (Fig. 4: V), although, its abundance remained
constant.

To summarize, on one hand, long-term coexistence of dominant populations is maintained in the *Caricetum gracilis* community, on the other hand, distinct dynamics of vegetation pattern and structure were found in the homogeneous environment occupied by
different facies of this community. Therefore the coincidence between life-history traits and
biotic interactions was investigated under a controlled water regime in the three dominant
species, in order to trace the establishment of the whole community.

Competitive ability of graminoid dominants

After three-year cultivation under terrestrial conditions the total biomass production
in monostands of *Carex gracilis*, *C. vesicaria*, and *Calamagrostis canescens* reached 1.74,
1.18 and 1.42 kg.m^{-2}, respectively. As the R/S ratio is much higher in sedges ($\approx$ 4) than in

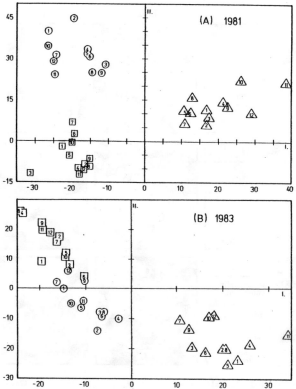

Fig. 3. PCA ordination of 36 permanent plots (1 × 1m) placed in different facies of the *Caricetum gracilis* community in 1981 and 1983. The abundance of species assessed by inclined point quadrat was ordered; I. and II. component intercept 32 and 36% of variability in 1981 and 1983, respectively; o — Car-grac-facies, □ — Cal-can-facies, △ — Car-ves-facies.

the grass (≤ 1.2), the aboveground biomass of *Calamagrostis canescens* is thus one third higher than in the most productive *Carex gracilis* stands. The aboveground maximum biomass in these experiments amounted to 0.34, 0.16 and 0.52 kg.m^{-2} in *Carex gracilis, C. vesicaria*, and *Calamagrostis canescens*, respectively, and corresponded with that assessed in the field, which in 1982 amounted to 0.44, 0.34, and .0.42 kg.m^{-2}, again respectively. In *Carex gracilis*, the biomass corresponds with that observed in dry biotopes of Central Europe, whereas in wetter biotopes the biomass is higher (Baradziej 1974, Hejný, Květ and Dykyjová 1981, Droste 1984).

The structure of the given stand within the observed community may therefore be determined by the interference of *Calamagrostis canescens* and *Carex gracilis* (*C. vesicaria* is the least productive species). Treatments focussed on their interspecific competition indicate that in terrestrial conditions *C. gracilis* is suppressed in the aboveground parts and consequently below ground (Fig. 5a). This suggests that strong competition is possible

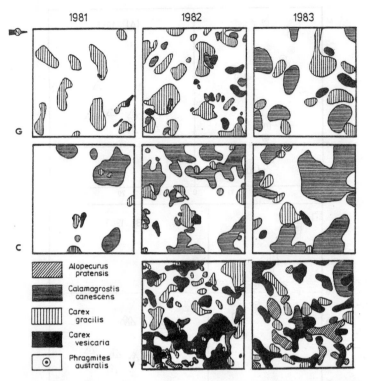

Fig. 4. Changes of microscale pattern within the Car-grac-facies (G), Cal-can-facies (C), and Car-ves-facies (V) illustrated by micromaps of $1 \times 1m$ large permanent quadrats in two or three successive years.

between the both species.

Rapid development of the sedge in spring and its intrinsic tillering confer only a weak advantage against the multiple impacts of the grass, of which shading is one of the most striking as culmless shoots of sedges are overgrown by 1.2 to 1.5 m tall tillers of the grass. This restriction is further enhanced by seasonal growth dynamics of the aboveground parts: the sedges reach their maximum biomass at the end of June, the grass attains the peak later — in middle July, which means that in July assimilation of storage materials in sedges is hampered by fully-developed grass tillers. Thus *Calamagrostis canescens* is the strongest competitor among the three species in terrestrial conditions, its dominance in Cal-can-facies on the driest sites within the community can be explained.

Higher water level inhibit the growth of all of the examined species, but to different degrees. Under limosal conditions the grass is markedly suppressed (its aboveground biomass falls to 0.33 kg.m^{-2}, and belowground organs — roots and rhizomes — produce only one third of the terrestrial biomass: 0.19 and 0.11 kg.m^{-2}, respectively), while in sedges only the roots are rather restricted. A remarkable decrease of seed and root production in

152

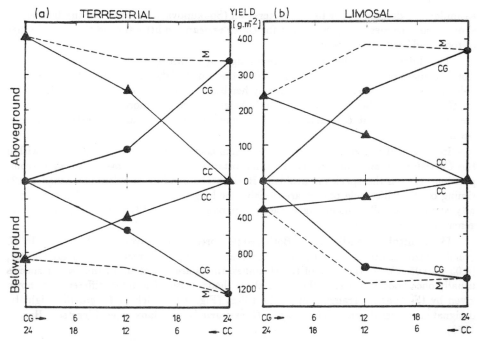

Fig. 5. Interspecific competition between *Carex gracilis* (CG) and *Calamagrostis canescens* (CC) under different water regimes expressed by de Wit's diagram (1960) for 3-year-old stands planted in sand; $\sum$ — total yield of the mixture.

sedges, was found only in littoral conditions, although aboveground biomass was highest under these circumstances — 0.40 and 0.21 kg.m^{-2} in *Carex gracilis* and *Carex vesicaria*, respectively. Adaptation of the sedges to waterlogging is related to their morphology and physiology e.g., aerenchyma (Metcalfe 1971, Soukupová 1986), special root surface, switches in respiration process (Končalová and Pazourek 1989). *Calamagrostis canescens* is unsuited to long-term persistence in inundation periods, its belowground parts are reduced to a third and its aboveground parts by a third. Thus its ability to compete is diminished and sedges instead of being eliminated become dominants. Long-term coexistence of *Calamagrostis canescens* with sedges is thus made possible by water level fluctuations.

The joint co-occurrence of the sedges with *Calamagrostis canescens* must, therefore, be classified as non-equilibrium coexistence. Micromaps of natural stands often record juvenile plants of the grass on the tops of sedge hummocks even on wetter sites. These hummocks provide two advantages for the grass: (i) due to centrifugal spreading the density of sedge shoots on top of the hummock is low thus making space available for germination and establishment of the grass and (ii) there are drier conditions on the hummocks.

Coexistence of the two sedges

Both sedges are adapted to waterlogging almost in the same way: The allocation of

153

assimilates is delayed under higher water level (Soukupová 1984), so that the biomass of shoots and rhizomes is higher (except for *Carex vesicaria* in littoral conditions) while that of roots and seeds becomes less.

The differentiation between the two sedges is shown by their morphological traits. Total biomass produced by *Carex gracilis* is one third higher than that of *C. vesicaria*. Additive features correlate as well, namely heights of plants: *C. gracilis* attains 1.2 to 1.5 m, *C. vesicaria* reaches only 0.7 to 1.0 m in height. Since it is more productive and taller, one might expect that *C. gracilis* would become strongly dominant in wetlands where it occurs.

In spite of this argument, stable-equilibrium coexistence of both sedges exists within the *Caricetum gracilis* community. Four important traits were discovered by which competition between the species is avoided. Horizontal spreading of tillers (1), allied to (2) timing of short life-cycle phases are decisive for long-term persistence within a community, structure of tillers including surface morphology of leaves (3) and (4) expenditure on generative reproduction.

Horizontal spreading. — Both sedges spread vegetatively by rhizomes. In *C. gracilis* rhizomes attain a length between 0.1 and 0.4 m, in *C. vesicaria* they are shorter, (0.03 to 0.05 m). At the ends of the rhizomes, tillers are initiated: In the former species usually one shoot appears, in the latter three shoots arise. The main difference is shown given by the spatial arrangement of the tillers (Fig. 6): which in *C. gracilis* might be designated as guerilla growth form while *C. vesicaria* is a phalanx type (Clegg sec. Harper 1978).

The difference was stated after three-year cultivations of plants from single vegetative shoots (Fig.6). In the former species 83 new tillers were established in clumps on several spots. In the latter, however, 480 tillers were formed side by side. The area occupied by these 3-year-old plants was the same in both species — 0.53 m². In *C. vesicaria*, the phalanx species, 70% of tillers were established between 0.1 m and 0.3 m from the parental tiller. *C. gracilis*, the guerilla species, formed two peaks of tiller density: the first between 0.00–0.15 m (28%) corresponded with the initial partial clump established from apogeotropic shoots around the parental tiller. The next more important peak (51%) occurred between 0.30–0.45 m and arose from longer diageotropic shoots (rhizomes).

Different intraspecific density of shoots is connected with their constitution. In the former species taller shoots are formed with higher number of leaves (± 6.1, in comparison with *C. vesicaria* with ± 5.0 leaves per shoot). Due to this attributes the former species occupies a territory only on several spots forming a basic network which is later saturated by the latter species.

Interlocked short life-cycles. — Sedge shoots experience closed development within periodically repeated short life-cycles (Soukupová 1988). This cycle includes the initiation of new shoots from buds of a parental vegetative tiller during primary tillering, and the growth of apogeotropic shoots until their full establishment as seasonal vegetative tillers. The transformation of shoots into generative tillers follows only after their overwintering. The duration of the short-life cycle is about two years in *Carex gracilis* and 18 months in *C. vesicaria*.

Substantial differences were found in the timing of primary tillering (autumn in *C.*

154

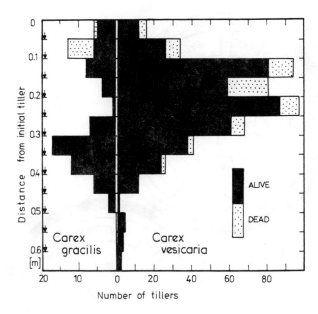

Fig. 6. Clonal spreading of tillers in 3-year-old plants of *Carex gracilis* and *C. vesicaria*, established from one initial (parental) tiller in a 1 × 1m plot.

gracilis, spring in *C. vesicaria*) and in the presence of overwintering tillers (*C. vesicaria* only). Delayed tillering enables *C. vesicaria* to produce new shoots only in free gaps amongst fully established tillers of *C. gracilis*, and a shorter assimilation period in summer is compensated by the occurrence of overwintering shoots which are important during temperate winters. (Fig. 7.)

Morphological adjustments. — *C. vesicaria* is appropriately equipped to fill its subordinate position. Its seasonal shoots 0.1 m shorter than those of *C. gracilis*, nevertheless the leaf area and weight of leaf blades is the same in both species. The difference in shoot length reflects the arrangement of leaves on the elongated basal part. In *C. gracilis* leaves of various lengths are formed, from shorter than 0.1 m to longer than 0.6 m, while *C. vesicaria* leaves are almost all of the same length, 0.3–0.4 m. In the former species shorter leaves are situated in the lower part of the growth axis while the longest reach to the canopy of the stand. In the latter species no pattern in vertical distribution of leaves was observed. Xeromorphic features in leaves of *C. gracilis* are advantageous for exposure to direct radiation (papillose epidermis, hypodermis, caespitose shape).

Generative reproduction. — *C. gracilis* produces twice the number of seeds (80,000 per m^2) as *C. vesicaria* (40,000 per m^2). Although germinability under full illumination of 520 W.m^{-2} is slightly higher in *C. gracilis* (34%) than in *C. vesicaria* (28%), in the field only seedlings of the latter species were common. In shade, germinability of *C. vesicaria* is higher compared to *C. gracilis* (Soukupová 1986: Fig. 33). Therefore seedlings of *C. vesicaria* establish themselves regularly in late spring every year within *Caricetum gracilis*

155

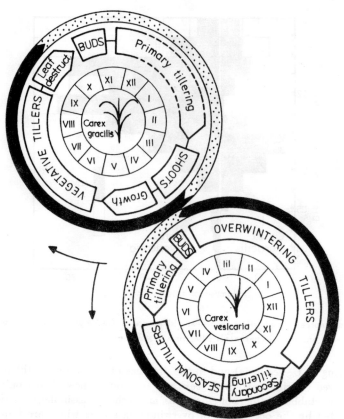

Fig. 7. Seasonal timing of growth in sedge tillers (developmental stages of vegetative organs are included; solid black — period with prevailing assimilation, dotted — period with prevailing clonal spreading).

in depressions with open canopy, while seedlings of *C. gracilis* were found only twice during the 10-year period of observations of the closed stands (in early summer of 1982 and 1989). This trait also contributes to successful survival of *C. vesicaria* within the closed stands of *C. gracilis*.

Conclusions

Development of spatial pattern within the *Caricetum gracilis* community results from adaptations and niche differentiation of the dominant species, their interference and interlocking of life-histories. As a result of these processes a model (Fig. 8) summarizing these processes can be presented with regard to the establishment of different kinds of stand within the community concerned.

Initial stands of *Caricetum gracilis* develop on regularly flooded biotopes, where regular occurrence of limosal or even of littoral conditions eliminates the seedlings of meso-

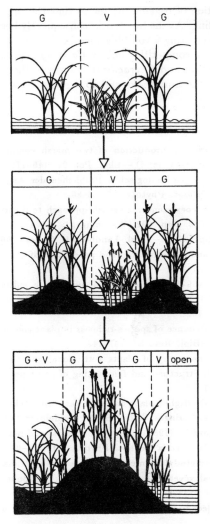

Fig.8. Scheme of pattern changes among dominant populations during the development of *Caricetum gracilis* community (G — *Carex gracilis*, V — *C. vesicaria*, C — *Calamagrostis canescens*).

phytic species. In this period young clumps of *Carex gracilis* shoots invade suitable sites in a guerilla way. *C. vesicaria* then penetrates in a phalanx way thus establishing in the open sites.

Later on when hummocks of *C. gracilis* are formed, *C. vesicaria* occupies the depressions which offer suitable spatial and light conditions. Terrestrialization proceeds during this period via accumulation of organic matter so that the upper horizon of hummocks escapes the effects of flooding. As the hummocks consequently become senescent, their upper parts are thinned thus creating favourable microhabitats for seedlings of *Calamagrostis canescens*. Due to its strong competitive ability *Calamagrostis canescens* quickly invades the top of the sedge hummocks and *C. gracilis* is suppressed there. The latter species then expands to lower parts of the hummocks and to depressions where it co-occurs with *C. vesicaria*.

References

Baradziej E. (1974): Net primary production of two marsh communities near Ispina in the Niepolomice Forest (southern Poland).—Ekol. Pol. 22: 145–172.

Blažková D. (1971): Zu der phytozönologischen Problemen der Association *Caricetum gracilis* Almquist 1929.—Folia Geobot. Phytotax. 6: 43–80.

Braakhekke W.G. (1980): On coexistence: a causal approach to diversity and stability in grassland vegetation.—Agr. Res. Rep. 902 (Wageningen): 1–164.

de Wit C.T. (1960): On competition.—Versl. Landbouwk. Onderz. 66: 1–68.

Droste M. (1984): Aboveground standing crop and production of *Carex gracilis* Curt. in a fen.— Arch. Hydrobiol. 100: 533–538.

Dykyjová D. and Véber K. (1978): Experimental hydroponic cultivation of helophytes.—In: Dykyjová D. and Květ J. [eds.], Pond littoral ecosystems. Springer Verlag, Berlin, Heidelberg, New York, p. 181–192.

Grubb P.J. (1977): The maintenance of species-richness in plant communities: the importance of the regeneration niche.—Biol. Rev. 52: 107–145.

Harper J.L. (1978): The demography of plants with clonal growth.—In: Freysen A.H.J. and Woldendorp J.W. [eds.], Structure and functioning of plant populations. Amsterdam, Oxford, New York, p.27–48.

Hejný S. and Husák Š. (1978): Higher plant communities.—In: Dykyjová D. and Květ J. [eds.], Pond littoral ecosystems. Springer Verlag, Berlin, Heidelberg, New York, p. 23–64.

Hejný S., Květ J. and Dykyjová D. (1981): Survey of biomass and net production of higher plant communities in fishponds.— Folia Geobot. Phytotax. 16: 73–94.

Hofman P. (1965): Jednotné metody chemického rozboru vod.—Praha.

Hroudová Z. (1988): Littoral vegetation of the Rožmberk Fishpond and its mineral nutrient economy.—Studie ČSAV 9/88.

Jeník J. and Květ J. (1983)[eds.]: Studie zaplavovaných ekosystémů u Třeboně.—Studie ČSAV 4/83: 1–156.

Kershaw K.A. (1973): Quantitative and dynamic plant ecology.—London.

Končalová H. and Pazourek J. (1989): Root dimorphism in flooded *Carex gracilis* Curt.—Acta Univ. Carolinae, Biologica 31: 43–50.

Lukavský J. (1975): Analysis of growth rate of algae by cultivation on solid media.—Arch. Hydrobiol./Suppl. 49, Algological Studies 14: 105–136.

Lukavský J. (1983): The evaluation of algal growth potential by cultivation on solid media.— Water Res. 17: 549–558.

Metcalfe C.R. (1971): Anatomy of the Monocotyledons V. *Cyperaceae.*—Oxford.

Orlóci L. (1975): Multivariate analysis in vegetation research.—Dr. W. Jung Publ., The Hague.

Přibáň K., Ondok J.P. and Jeník J. [in prep.]: Microclimate in the Třeboň Biosphere Reserve.—Studie ČSAV.

Soukupová L. (1984): Production ecology in a sedge-grass marsh.—In 2nd INTECOL Wetland Conference, Třeboň, Czechoslovakia 13–23 June 1984, Abstracts I: p. 106.

Soukupová L. (1986): Studie životnístrategie u mokřadních travin.—Ms. [Thesis, Institute of Botany, Třeboň].

Soukupová L. (1988): Short life cycles in two wetland sedges.—Aquat. Bot. 30: 49–62.

Warren Wilson J. (1959): Inclined point quadrats.— New Phytol. 59: 1–7.

Krahulec F., Agnew A.D.Q., Agnew S. & Willems J.H. [eds.]: *Spatial processes in plant communities.* pp. 161–172.

Bryophyte pattern and dynamics in Dutch chalk grassland

Hans van Rijnberk[1,2] and Heinjo During[1]

1) *Department of Plant Ecology, University of Utrecht, Lange Nieuwstraat 106, 3512 PN Utrecht, The Netherlands*
2) Present address: *Department of Evolutionary Biology, University of Leiden, Schelpenkade 14A, 2313 ZT Leiden, The Netherlands*

Keywords: Transition matrix, Mowing, *Pseudoscleropodium purum, Ctenidium molluscum, Fissidens taxifolius, Calliergonella cuspidata, Eurhynchium hians.*

Abstract. Point-quadrat grid recordings were taken in March and May 1983 of the bryophyte layer in four dutch chalk grasslands. For each species pattern was analysed using correlograms based on join-counts. Pattern was assumed to correlate with dynamics, which was calculated as the relative cover change between recording dates within grids of each grassland. Small-scale dynamics, defined as the number of replacements of species i (March) by species j (May), were tested for directional changes in species composition.

For pattern, general trends show clustering on a small scale, gradually fading with increasing distance. Differences mainly occurred in clustering intensity and to a lesser extent in the rate of its decline over distance. There was no relation with grassland type or life strategy.

Contrary to expectation, dynamics did not correlate with pattern. Small-scale dynamics, inferred from interspecific transition matrices, showed a site-specific succession order for species. This indicates directional changes in species composition, governed by either abiotic influences or species interactions (competition) or both. The results allow the formulation of specific hypotheses for future experiments.

Introduction

Species pattern in any vegetation both reflects and affects the ecological conditions under which the plants are growing. Often it co-varies with dynamics, which itself can be described as the path of pattern through time. Gradually more knowledge is being gained about both aspects (Greig-Smith 1979). Correlations between them may yield information on processes, although experiments are needed for proof (e.g., Goldberg and Werner 1983, Yodzis 1988). For the study of a process one needs to make a choice on the subject, the environmental conditions and, most importantly, on the hypotheses or assumptions about the nature of the process (Goldberg and Werner 1983). Exploratory research on the relation between pattern and dynamics should contribute to a sound basis for such assumptions. A suitable model to start from is that of Lovett Doust (1981, 1987, Lovett Doust

161

and Lovett Doust 1982) in which species are arranged in a Phalanx-Guerilla continuum. Pure Phalanx species would invade a habitat from a tight front and keep it occupied for long time. Pure Guerilla species would wander through the vegetation; as a consequence the shoots are rather short-lived. Thus the Phalanx strategy correlates with clumping and a low dynamics contrary to the Guerilla strategy.

Bryophytes are very suitable objects for research on pattern and dynamics:
- the pattern is mainly two-dimensional;
- the plants lack roots (no below-ground processes); and
- the species form a good representation in the Phalanx-Guerilla continuum as they show a wide range of growth forms and reproductive modes.

Dutch chalk grasslands are very rich in bryophytes. In view of the lack of roots, a height of some cm at most and very little seasonal differentation, it may be presumed that the coexistence of species is mainly based on gap-dynamics (intermediate disturbance hypothesis, Huston 1979, During and van Tooren 1987, 1988). As pleurocarpic species dominate, in which recruitment is nearly exclusively by horizontal growth and branching, pattern will probably be ruled by clustering. Differences between species in scale and intensity of clustering are likely to occur (During and van Tooren 1988). The latter will determine the degree of contact between species as well as the impact of the environment. Thus dynamics are predicted to be negatively correlated with cluster intensity, differentiated by the cluster-scale. Small-scale dynamics imply changes in species composition. If reduced to point-observations as the smallest scale possible, such changes can supply information on species relations.

Methods

Sites

Sites were chosen so as to differ in exposition and management but to be similar in other respects. Four sloping grasslands were selected in South-Limburg (The Netherlands) of which two were mainly North- and two South-exposed; one of each pair was grazed by sheep and one annually mown. The North-facing sites are adjacent (Gerendal) and were mown annually until 1979 after which one was managed by sheep grazing. The South-facing sites are more different from each other. The Vrakelberg had been mown for many years before, the Bemelerberg had been abandoned untill 1980; in that year a grazing management regime was installed. This site had a very low bryophyte cover and contained only one quantitatively important species.

Species composition differed among sites (Table 1). If species occurred in more than one site, the abundances differed markedly, which may have implications for the analysis of pattern.

Sampling

Because analysis based on sampling of distances between moss individuals in such a dense and species-intermingled community is impossible, point-interception quadrat sampling was applied. In spring 1983, seven randomly chosen plots of 25 × 25cm were recorded twice at each site, once in March and once in May. Each plot consisted of a grid of 19 × 19

162

Table 1. Cover of the quantitatively most important species; mean percentage per plot and standard deviations. (N – north, S – south, G – grazed, M – mown; P – perennial stayer, C – colonist, and L – longlived shuttle.)

species (strategy)	date	sites							
		N–G		N–M		S–G		S–M	
		mean	sd	mean	sd	mean	sd	mean	sd
Calliergonella	March	20.8	22.6	32.4	28.7			0.1	0.2
cuspidata (P)	May	22.2	33.6	31.2	37.1			0.2	0.4
Campthothecium	March	0.0	0.0	0.2	0.3			0.2	0.4
lutescens (P)	May	0.0	0.0	0.1	0.3				
Ctenidium	March	32.2	29.8	10.3	13.0				
molluscum (P)	May	29.6	28.5	14.9	15.8				
Eurhynchium	March	3.6	5.2	1.9	5.2	15.6	21.5	1.3	2.3
hians (P)	May	2.5	3.1	2.3	3.3	6.9	7.8		
Eurhynchium	March	0.2	0.5	1.6	2.4	0.4	0.8	11.7	20.0
striatum (P)	May								
Fissidens	March	4.9	3.4	6.2	10.2			2.5	4.2
taxifolius (C)	May	6.0	5.1	5.8	8.7			2.3	2.7
Fissidens	March	0.0	0.1					2.5	4.7
cristatus (C)	May							1.0	2.9
Plagiomnium	March	18.7	33.7	11.6	11.5				
undulatum (L)	May	17.6	32.9	9.2	8.5				
Pseudoscleropodium	March	3.8	5.5	1.4	1.4			5.0	10.8
purum (P)	May	4.1	5.5	5.8	8.6			7.7	15.5
Rhytidiadelphus	March	0.4	1.0	2.6	5.1			0.0	0.0
squarrosus (P)	May	0.0	0.0	2.8	5.2			0.0	0.0
Brachythecium	March	0.0	0.0	0.0	0.0			2.0	2.9
glareosum (P)	May	0.0	0.0	0.0	0.0			2.0	2.9
Weissia	March					1.5	2.1	1.4	2.4
controversa (C)	May					0.6	1.3	2.3	2.7

points (interpoint distance = 1.25 cm). Species presence or absence for every point location was noted.

Pattern analysis

All the grids were analysed for each species separately by join-count analysis (Upton and Fingleton 1985).

Joins are neighbouring observations with the same species, here counted separately for each scale of 1 to 9 cells distance in horizontal and vertical direction (rook's case). The results thus obtained form species-specific join-count correlograms when plotted as join-counts versus distance. Expected values and standard deviations under the null hypothesis of random pattern were obtained using the formulæ in Upton and Fingleton (1985). The join-counts were transformed to Z-values ($Z = (O - E)/SD$), which follow the standard normal distribution for random pattern. The graphic results obtained is called a Z-correlogram (Z-value over distance). Pattern on a certain scale (d) is considered to denote clustering if $Z > 1.96$ and regularity if $Z < -1.96$. In the few cases for which the expected number of join-counts was ≤ 3 the Poisson distribution was used to test significance and the Z-correlogram was calculated using the square root of the expected value as an estimate for the standard deviation under randomization.

The resulting dataset of about 170 correlograms was analysed by K-means clustering (Sokal and Thomas 1987; SAS FASTCLUS procedure, 1985) for overall structure. The method is very flexible and therefore does not suffer from misclassification as much as e.g. with agglomerative hierarchical clustering of correlograms (see Sokal and Thomas 1987). Observations on species with less then 5 points in a grid or with a cover of more than 75% (gridpoints) were discarded. Preliminary analysis showed that cover influences the results severely in these cases. Pattern then is physically limited which for the latter case implies that pattern occurred on a larger scale. Cluster differences were visualised with the help of Canonical Discriminant Analysis (SAS CANDISC-procedure, 1985). It creates a reduced number of variables which are linear combinations of the original variables (Zd) maximizing cluster differences. The resulting scattergrams were inspected for correlations with life-strategy and site-characters by attaching labels to data-points in the graph. Site labels were north-grazed—1, north-mown—2, south-grazed—3 and south-mown—4. Life-strategies distinguished, following the nomenclature of During (1979), were perennial stayers (P), colonists (C), long-lived (L) and annual shuttle species (A).

Relative cover change

For every species in a recording the relative change from March (1) to May (2) was calculated as $(N_2 - N_1)/(N_2 + N_1)/2$, with N as the number of point observations in a grid.

Small-scale dynamics

For each site a March to May transition matrix was assembled by counting the transitions from species i to species j. Species with low cover were pooled in a separate class called 'grain species' for convenience. Open space was treated as a separate species. For

164

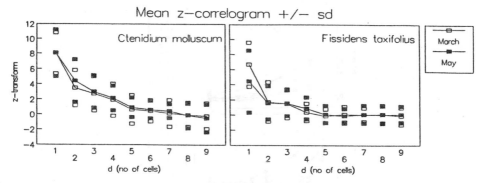

Fig. 1. Mean correlogram with standard deviations for *Ctenidium molluscum* and *Fissidens taxifolius* in March (open squares) and May (filled squares).

the south-grazed site the matrix contained too many zero transitions and it has therefore not been analysed.

Analysis was performed using Iterative Sequential Testing of Adjusted Residuals (IS-TAR; Haberman 1973, van Hooff 1980), which calculates Chi-square and independent z-values (adjusted residuals) for every cell (= transitional element). The procedure tests the overall significance of the transition-structure by a normal Chi-square test for contingency tables. If the structure is non-random it determines which significant transitions (= cells) in rank order contributed most to the structure. The results show species-specific replacements by other species.

A positive sign of a significant Z-value indicates that species i is replaced by species j more frequently than expected on the basis of their abundance (in March and May). A negative value indicates that this occurs less frequently than expected. For the south-mown site both a matrix of eight species (of which one was a 'grain species') and a matrix of four species (with the category 'grain species' including the other four species) were analysed. Testing was performed using respectively a 2.5% and 5% significance level. This special treatment was applied to avoid testing errors because of the high number of cells (40%) with expected frequencies < 5 (Haberman 1973, van Hooff 1980, Siegel 1956). Our null-distributions (Chi-square and Standard normal) are therefore less valid, as this number should not exceed 25% of the matrix cells (Haberman 1973, van Hooff 1980, Siegel 1956). Testing of the separate cells however is far less sensitive in this respect than the Chi square test (previously cited authors). The special treatment as well as the high significance levels therefore still enable meaningfull interpretation for these results.

Results

Pattern of species

Pattern within species varies considerably. As a general trend, species show short-distance clustering gradually fading away with distance to zero (Z). This implies that there is no cluster size to be distinguished. Then the correlogram should show a positive significant Z-value (coherence within the cluster) with increasing distance followed by a

165

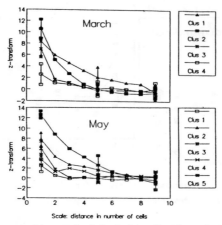

Fig. 2. Mean correlograms with standard deviations (for three d values) of the March and May
clusters.

negative significant value (contrasting with the background). An illustration of the general
features of species pattern is given in Fig. 1 by mean correlograms for two species. *Ctenidium molluscum* as a densely branching pleurocarp with thick mats can be regarded as a
phalanx species; *Fissidens taxifolius,* an erect acrocarpic species with small turfs is considered to represent a guerilla strategy. These differences become apparent by the steeper
decline of Z over distance for the latter.

Pattern over all species

Overall structure in pattern can be characterised by four clusters in March and five in
May (Fig. 2). Mean cluster-correlograms display typical differences between the clusters,
mainly determined by the cluster-intensity at distance 1 and to a lesser extent by the rate
of decline in clustering over distance.

The canonical variables resulting from the CANDISC procedure are strongly correlated with small-scale pattern; nearly all variation is contained within the first two variables
(Table 2). Visual inspection from CANDISC plots with either life-strategy or site number as datalabels (Fig. 3) does not show any relation with the cluster characterisation of
pattern.

Relative change in cover

Variation in relative change in cover decreases with increasing cover (Fig. 4). Thus,
locally frequent species tend to constancy, whereas dynamics of locally sparse species may
be either low or high. This probably indicates that the time-scale of dynamics differs more
for low cover observations than for high ones. A plot of relative change in cover (dynamics)
against Z1, as being the most important determinant of cluster-differences, reveals no
relationship (Fig. 5); which contradicts the hypotheses on the negative correlation between

166

Table 2a. Squared correlations from CANDISC analysis on clusters of species pattern correlograms.

	Z_1	Z_2	Z_3	Z_4	Z_5	Z_6	Z_7	Z_8	Z_9
March	.73	.69	.46	.38	.34	.41	.29	.27	.13
May	.85	.80	.76	.61	.43	.21	.13	.09	.18

Table 2b. Squared canonical correlations with explained percentage of total variation.

	CAN 1	CAN 2	CAN 3	CAN 4
March	.82 70.0%	.62 25.6%	.22 4.5%	
May	.91 87.0%	.57 11.0%	.12 1.1%	.10 0.9%

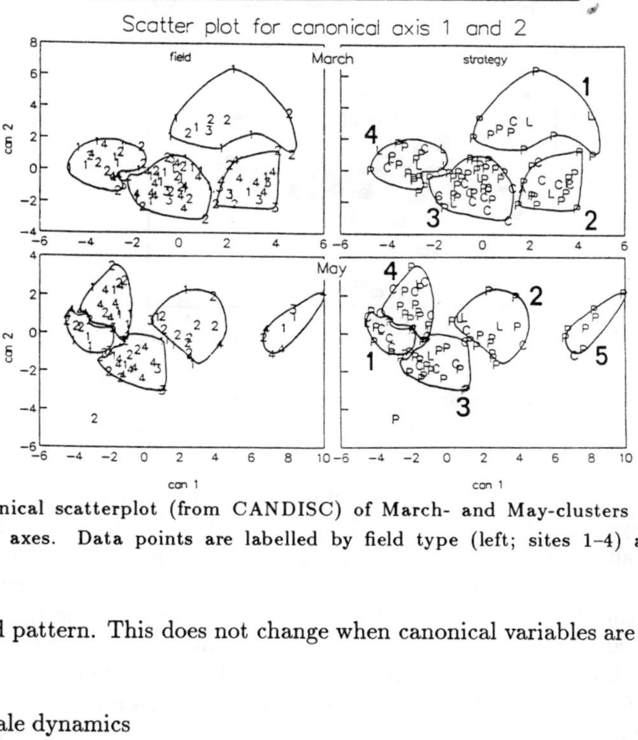

Fig. 3. Canonical scatterplot (from CANDISC) of March- and May-clusters for the first two canonical axes. Data points are labelled by field type (left; sites 1–4) and life strategy (right).

dynamics and pattern. This does not change when canonical variables are used instead of Z1.

Small-scale dynamics

The three sites analysed by ISTAR showed a highly significant transition structure (Table 3), indicating a directional change in species composition over the sampling period. Tests on separate cells or elements of the transition matrix reveal which species transitions contributed most to the dynamic structure. To make inferences about species relations it

Bryophyte dynamics related to cover

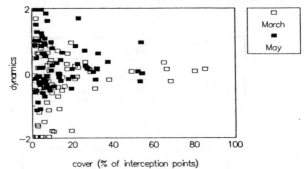

Fig. 4. Scatterplot of dynamics against percentage cover calculated from gridpoint counts in March (open squares) and May (filled squares).

Bryophyte dynamics related to Z1

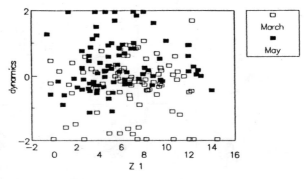

Fig. 5. Scatterplot of dynamics against Z1 for the March (open squares) and May recordings (filled squares).

is necessary to distinguish different types of elements. A significant 'diagonal element' of a species (always positive) denotes that it shows less change than expected and therefore seems to be locally stable within plots. When an element denoting a replacement of species i by species j $(i \neq j)$ is compensated by the reverse element in the same matrix, it is called a 'reversible element'. The matrix represents a site and therefore the species then seems globally stable among plots within the site. If elements are not compensated they are denoted as 'unidirectional'. An example for the North-grazed site is given in Fig. 6. All such elements for a site, combined in a scheme as in Fig. 7, show how settlement places are transferred among species. The number of elements for a site is expressed relative to the maximal possible number for diagonal, reversible, unidirectional and all elements to compare the sites.

The percentage of all elements being significant is highest at the north-mown site, lowest at the south-mown site. A high percentage of significant diagonal elements is ob-

Table 3. Chi-square test and number of significant transitional elements (sig) resulting from the ISTAR procedure. Also indicated: maximum possible number of transitional elements (max) and the % of those that are significant. (N – north, S – south, G – grazed, M – mown; information on S–M is from combined 8×8 and 4×4 ISTAR, leaving the exact number of diagonals in doubt!)

site	χ^2	df	total			diagonal			reversible			unidirectional		
			sig	max	%	sig	max	%	sig	max	%	sig	max	%
N–G	1643	36	11	49	23	4	7	57	0	42	0	7	21	33
N–M	1232	36	21	49	43	7	7	100	10	42	24	5	21	24
S–M														
8x8	985	49	13	64	20	7	8	88	2	12	17	2	6	33
4x4	792	9	7	16	44	4	4	100	1	56	2	4	23	17
tot			14	64	22	7/8	8	88	1	56	2	5	23	22

```
                    May species
          March       1      2      3      4      5   grain  blank
          species
             1     14.79     .      .      .      .      .      .
             2        .    21.91    .     3.84    .      .      .
             3        .      .      .      .      .      .      .
             4        .      .    -2.74  28.37    .      .    2.61
             5        .     5.89   2.85    .      .      .      .
          grain       .      .    -1.87   3.66    .     6.60    .
          blank       .      .    -2.69    .      .      .      .

          X²= 1643, df= 36

   SIGNIFICANT TRANSITION ELEMENTS BY TYPE

      * total                             : 11 out of 49, 23%;
      * diagonal (± local stability)      :  4 out of  7, 57%;
      * reversible (± global stability):     0 out of 42;  0%;
      * unidirectional (competition ?)  :    7 out of 21; 33%;
```

Fig. 6. ISTAR transition matrix of the bryophyte layer of the north-grazed site. Sequential testing of cells was continued until the first non-significant element (grain/spec. 3) was encountered.

tained at both mown sites and a relatively low one at the north-grazed site. Reversible elements were high in number at the north-mown site only.

The relative contribution of significant uni-directional elements was highest at the north-grazed site, indicating a strong hierarchy. Figure 7 gives the hierarchy abstracted from these elements for each site. The one south-facing site differs from the others by the central role of open space which directly relates to the relatively low bryophyte cover (1.7–57% compared to > 43% for the two north-facing sites). It consists of a split hierarchy in contrast with the north facing sites with a top-down hierarchy.

Fissidens taxifolius, a small erect moss, occurs in all sites but differs in hierarchy position. In the south-mown and the north-grazed schemes it replaces *Eurhynchium hians* (small appressed mats) but in the north-mown scheme it is replaced by all other contributors. In both north-facing sites it is replaced by *Plagiomnium undulatum*, the latter being

169

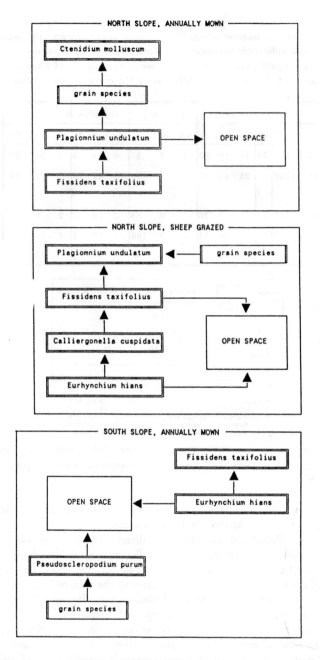

Fig. 7. Hierarchy schemes resulting from ISTAR 'unidirectional elements' for 3 of the sites.

at the top at the grazed site but more at the bottom at the mown site. In view of its growth form, the dominance of *Fissidens* is not likely to be governed by competition; its suppression however may be.

In the north-mown scheme 'grain species' constitute an element in the middle of the hierarchy, whereas in the other schemes this category occupies 'bottom positions'. In all schemes 'open space' occupies 'top-positions', relating to a net decrease in bryophyte cover.

Discussion

All quantitatively important species show some degree of clustering, at least on a small scale. With increasing distance this decreases gradually without showing distinct clusters. The bryophyte layer thus appears to be a network of clustering centers from which lateral spreading can take place over shorter or longer distances. Different types of pattern can be distinguished although type differences cannot be explained easily.

During and van Tooren (1988), using a grid of the same size as ours but with cells as sampling units instead of points to sample chalk grassland bryophytes, found a combination of clumping and fine grain. Spatial clusters of the different species seemed to overlap. Fine grain was mainly observed within cells, a scale for which the present study does not supply data. They suggest that the bryophyte layer consists of 'guilds' of species with similar life strategies; interactions between species were only found within such guilds (cf. Shmida and Ellner 1984, Grubb 1986). However, the present analysis on transition structure suggests that a species does not have the same ecological function in every site. Why else does a species show different behaviour over sites? Differences in cover within one species vary considerably among the sites, resulting in scale differences of the pattern. This may explain why generalised species or life-strategy patterns are not present. Site- (or grid-) specific patterns on the other hand could be feasible as shown from the site-specific transition structure and species relations.

During and van Tooren (1988) found a positive correlation between vegetation pattern and dynamics with whole grids as units. This was interpreted as indicating intermediate disturbance causing high species diversity and therefore a high degree of intermingling. Thus on the community level within grid units, their data seemed to accord with our predictions, although they were not interpreted in the context of our hypothesis. Temporal processes over the sample period suggest directional changes in the species composition. The resulting species transitions proved to be structured and site-specific. Furthermore a site-specific sequence of species was demonstrated suggesting the presence of driving processes. We suggest that environmental and/or competition processes play a role. Suitable objects and relational hypotheses for experiments can be chosen from the analysis but should be site-specific. For instance the replacements by *Fissidens* (see results) can be tested by changing environmental factors and/or gap creation (van Tooren *et al.* 1987, During and van Tooren 1988). The relation with the species to which it appeared inferior could be tested for competition. Furthermore *Ctenidium molluscum, Eurhynchium hians* and *Calliergonella cuspidata* seem suitable objects for future experimentation as long as specific conditions are kept the same. Competition is therefore best tested by pulse perturbations (Bender *et al.* 1984) in the natural environment.

171

References

Bender E.A., Case T.J. and Gilpin M.E. (1984): Perturbation experiments in community ecology: theory and practice.—Ecology 65: 1–13.

During H.J. (1979): Life strategies of bryophytes: a preliminary review.—Lindbergia 5: 2–18.

During H.J. and van Tooren B.F. (1987): Recent developments in bryophyte population ecology.— -Trends Ecol. Evol. 2: 89–93.

During H.J. and van Tooren B.F. (1988): Pattern and dynamics in the bryophyte layer of a chalk grassland.—In: During H.J., Werger M.J.A. and Willems J.H. [eds.], Dynamics of pattern and diversity in plant communities. pp. 195–208. SPB Academic Publishing, The Hague.

Goldberg D.E. and Werner P.A. (1983): Equivalence of competitors in plant communities: a null hypothesis and an experimental field approach.—Amer. J. Bot. 70: 1098–1104.

Greig-Smith P. (1979): Pattern in vegetation.—J. Ecol. 67: 55–779.

Grubb P.J. (1986): Problems posed by sparse and patchily distributed species in species rich plant communities.—In: Diamond J. and Case T.J. [eds.], Community Ecology, pp. 207–226. Harper and Row, New York.

Haberman S.J. (1973): The analysis of residuals in cross-classified tables.—Biometrics 29: 205–220.

Huston M. (1979): A general hypothesis of species diversity.—Amer. Nat. 113: 81–101.

Lovett Doust L. (1981): Population dynamics and local specialisation in a clonal perennial (*Ranunculus repens*) I. The dynamics of ramets in contrasting habitats.—J. Ecol. 69: 743–755.

Lovett Doust L. (1987): Population dynamics and local specialisation in a clonal perennial (*Ranunculus repens*) III. Responses to light and nutrient supply.—J. Ecol. 75: 555–568.

Lovett Doust L. and Lovett Doust J. (1982): The battle strategies of plants.—New Scientist 95: 81–84.

SAS Institute Inc. (1985): SAS/STAT quide for personal computers, version 6. Edition. Cary N.C.: SAS Institute Inc.

Shmida A. and Ellner S. (1984): Coexistence of plant species with similar niches.—Vegetatio 58: 29–55.

Siegel S. (1956): Non-parametric statistics for the behavioural sciences.—Chapt. 9, Contingency analysis, pp. 196–202.

Sokal R. and Thomas J.D. (1987): Applications of spatial autocorrelation in ecology.—In: Legendre P. and Legendre L. [eds.], Developments in Numerical Ecology. Springer Verlag, Berlin.

Upton G.J.G. and Fingleton B. (1985): Spatial data analysis by example I. Point pattern and quantitative data.—Wiley and Sons, Chichester.

van Hoof J.A.R.A.M. (1980) Categories and sequences of behaviour; methods of description and analysis.—In: Scherer P.G.K. [ed.], Handbook of non-verbal communication research, chap. 6. Cambridge University Press, New York.

van Tooren B.F., During H.J. and Oudhof J.A.F. (1987): Dynamics of bryophytes in chalk grasslands.—Folia Geobot. Phytotax. 22: 377–383.

Yodzis P. (1988): The indeterminacy of ecological interactions as perceived through perturbation experiments.—Ecology 69: 508–515.

Krahulec F., Agnew A.D.Q., Agnew S. & Willems J.H. [eds.]: *Spatial processes in plant communities.* pp. 173–184.

Fine scale dynamics in a mountain grassland

Tomáš Herben, František Krahulec, Marcela Kovářová, and Věra Hadincová

Botanical Institute, Czechoslovak Academy of Sciences, 252 43 Průhonice, Czechoslovakia

Keywords: Species replacement, Permanent plots, Manuring, *Nardus*-grassland, Species competition, Interaction variation

Abstract. Four permanent plots (50 × 50 cm) of a *Nardetalia* grassland (Krkonoše Mts., part of the Sudeten Mts.) with only five important species per plot were recorded yearly from 1984 to 1989. Each plot was divided into 225 subplots, where number of vegetative units (tillers for grasses, leaves for dicots) were counted for each species. Two plots were manured in 1985.

Each species had a specific temporal pattern in frequency, density in subplots and stability in space within the plot. Frequency was much more species specific, while density reflected mostly the management regime.

Small scale dynamics were expressed as replacements of component species at the level of subplots. The direction and presence of replacements between species varied both in time and between plots; unidirectional replacements were found only in manured plots. Interspecific associations between the species were not related to their dynamic relation (direction and strength of replacement). *Deschampsia flexuosa* and *Anthoxanthum alpinum* showed low spatial stability in time and were important mainly in short term replacements (1 year); *Nardus* showed high spatial stability and was important in long term replacements (over 5 years). Great variation between plots was found both in interspecific associations and replacement patterns.

Introduction

There is still little known about the competitive interactions of wild plants under natural conditions. Strictly speaking, competition cannot be addressed without some experimental disturbing of the community. However, the crucial role of spatial processes in plant competition allows us to approach competition through the study of spatial replacement processes, which in many plant systems may be reasonably thought of as a direct outcome of the competition (though this has to be proved independently). Such studies are most easily done in two dimensional systems (like lichens, bryophytes or, among animals, sessile invertebrates), where pairwise interactions can be singled out of the community processes and studied separately (Hawksworth and Chater 1979, van Rijnberk and During 1990). By no means is this relatively straightforward approach limited to these particular plant groups. It can also be applied to some higher plant systems, given that the spatial resolution in the data collecting process is fine enough. Such an approach would theoretically allow one to construct spatial replacement diagrams between species pairs, which could possibly be compared with independently obtained data on competition of the species. Obviously, care should be taken in the interpretation of the results, which should preferably be called replacement rather than competition.

In grasslands, the dynamic relations between species are important especially in determining the species composition and abundance. There are plenty of data sets on the overall (macroscopic) dynamic behaviour of grassland swards (e.g. Grubb *et al.* 1982, Rabotnov 1984), but there are surprisingly few data on the dynamic processes at a fine spatial scale (e.g. species replacements, species mobility as related to community structure etc.). Knowledge of these fine scale dynamics is important in two respects: (1) it allows one to understand the macroscopic behaviour of the system and is helpful in making nontrivial predictions, and (2) it is (given a homogeneous space) related to spatial pattern of the community (sociological pattern *sensu* Kershaw 1973).

We started a "high-resolution" permanent plot study in a mountain grassland to get an insight into the internal dynamics of the grass sward. The great advantage of this particular grassland type is twofold: (1) very low species number, which allows the study of all possible species pairs, and (2) a moderate rate of dynamic behaviour due to the harsh climate. Our principal aims were to investigate:

1. Movement vs. stability of species, and rate of their invasion of space;
2. dynamics of species interactions (species replacements, and their stability in time and space) and their relation to interspecific associations;
3. effect of fertilization on the above processes.

Methods

Study site

The data were collected in the mountain grassland in the Krkonoše Mts., NW part of Czechoslovakia (Severka settlement, ca. 3 km NW of Pec pod Sněžkou, altitude approx. 1100 m a.s.l.—see Fig.1), at a meadow managed in the traditional way. These meadows were mowed once or twice a year (depending on the climate) and grazed late in the autumn. They were manured once in several years. The studied plots are rather species poor, with only five important species: *Anthoxanthum alpinum*, *Deschampsia flexuosa*, *Festuca rubra*, *Nardus stricta* and *Polygonum bistorta* (At some plots, *Vaccinium myrtillus* also was present). These species together always accounted for 95% of the total living aboveground biomass. From the phytosociological point of view, the meadows can be classified into *Sileno-Nardetum pleurozietosum* (*Nardo-Agrostion* alliance, *Nardetalia* order)(Krahulec 1990).

Data collection

Four permanent plots of 50 × 50 cm were established (three in 1984 and one in 1985) and recorded once a year (ca. mid July) till 1989. The recording was made in the following way: a grid of 15 × 15 subplots (i.e. each subplot was 3.3 × 3.3 cm in size) was established at the plot and the number of vegetative units of all plants within each subplot was counted. For grasses, these units were defined as tillers (ramets); for forbs (primarily *Polygonum bistorta*), units were defined as individual leaves. Further, the plots were clipped after recording, the clippings being used for estimation of the living aboveground

174

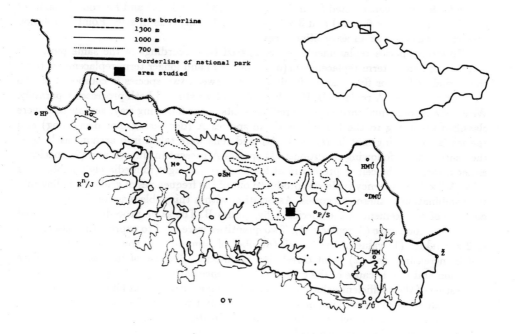

Fig. 1. Location of the study area

biomass of individual species on the whole plot (except in 1985). Two plots (S1 and S3) were manured in autumn 1985 (using cow dung, pH=7.7, which amounted to adding the following amounts of nutrients in $g.m^{-2}$: 17 N, 0.2 nitrate-N, 3.8 ammonia-N, 2.4 phosphate-P).

Data analysis

The data were processed to yield information on a) global dynamics of individual species in the whole plot, b) fine scale dynamics of individual species, and c) species interactions at the small scale.

For each plot in each year the following global quantities were calculated: a) above-ground biomass of each species, b) total biomass as a sum of component species biomasses, c) relative biomass of each species, i.e. proportion of the total biomass, d) species frequency, i.e. number of cells occupied by the species, e) species density, i.e. mean number of units of the species in occupied cells.

The fine scale spatial dynamics of an individual species was expressed by correlating the initial value of the species at a given cell with that species value at different time lags. This produced a "decay curve" of each species at the level of the cell size. Spearman rank correlation coefficient was used.

Interspecific associations were studied by means of 2×2 presence/absence tables. These tables were constructed for individual cells (3.3×3.3 cm) and for combinations of cells 2×2 (i.e. 6.7×6.7 cm) and 3×3 (i.e. 10×10 cm) cells in size. Point correlation coefficient was used as an association measure.

To study species replacements, we compared two recordings of the same plot. We studied both short term replacements (over one year) and longer term replacements (over the period of four or five years). Replacement between two species is composed of two separate processes (A replaces B, B replaces A) and we studied both processes separately. We estimated only qualitative species replacements in the following way: sample cells were classified according to the behaviour of the species A (decreasing, not decreasing) and species B (increasing, not increasing). All grid cells were taken into account as long as they satisfied the following conditions: (1) species A present in the cell and (2) species B either present in the cell or present in at least one of the eight immediately neighbouring cells. A 2×2 contingency table was produced out of the frequencies of grid cells falling into the combination of these categories. Association within this table was considered to be a measure of replacement of the species A by species B. The same procedure was repeated for the reciprocal case (A replaces B). Two quantities were used to express this association in 2×2 table: (1) value of the 'a' cell (number of common occurrences of A decreasing and B increasing, and (2) point correlation coefficient. Because of the nonrandom data collection and also because of the complex dependence within the data set, we did not do any testing of the coefficient values; instead, we set an arbitrary limit for each coefficient to delimit values we considered important. The same limit was used for each coefficient in all data sets. The limit was set so as to cover approximately one quartile of the distribution of the observed values for each coefficient in each data set (one plot in static analyses, plot-plot in dynamic analyses). For point correlation coefficient, 0.15 was used; for the 'a' cell, 10 was used in short term comparisons and 15 in long term comparisons. Only species pairs with more than 20 observations (grid cells) per plot were considered. This led to the exclusion of the following species pairs: *Nardus–Festuca* (Plot 2, 4), *Nardus–Polygonum* (Plot 2).

Results

Dynamics of the whole plot

Total biomass of the plots varied considerably (by up to a factor of two, Fig. 2); the variation between years was greater than variation between plots. The same variation was observed in biomasses of individual species (resulting from variation of the vegetative unit mass). Variation among plots increased in the course of the observations. Proportions of biomass of individual species differed between manured and unmanured plots (Fig.3): in manured plots, the proportion of *Festuca* increased during the whole period of observation, the proportion of *Anthoxanthum* only up to the 2nd–3rd year after manuring, whereas proportions of *Deschampsia* and *Polygonum* decreased considerably. In unmanured plots, the biomass proportions were more constant and did not show clear trends during the observation period.

Changes in frequency and density of species are summarized in Table 1. *Nardus stricta* and *Deschampsia flexuosa* increased in density in unmanured plots, while in manured plots

176

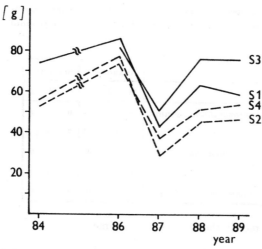

Fig. 2. Changes of the total living aboveground biomass for individual plots. In 1985, biomass was not estimated. Plots S1 and S3 were manured in 1985.

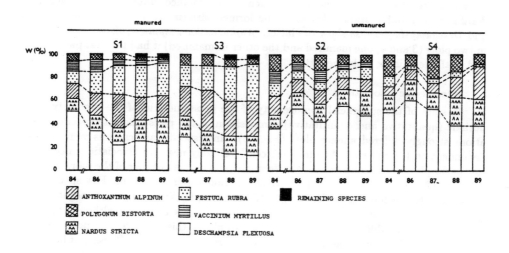

Fig. 3. Changes in living aboveground biomass proportion of individual species.

this trend was not so obvious. *Festuca rubra*, *Anthoxanthum alpinum* and *Polygonum bistorta* increased in density only in manured plots and no trends were found in unmanured plots. Frequency of *Anthoxanthum* increased as did density in manured plots; for other species no connections with density was found. Frequency was much more species-specific,

177

Table 1. Trends in global species responses. W—relative biomass; d—density; f—frequency; +—increasing; −−decreasing; 0—no change.Doubled label indicate pronounced change, label in brackets indistinct changes.

	manured			unmanured		
	W	d	f	W	d	f
Anthoxanthum alpinum	+	+	++	0	0	+
Deschampsia flexuosa	−	(+)	(−)	0	++	(−)
Festuca rubra	++	++	+	0	+	0
Nardus stricta	0	+	0	+	++	+
Polygonum bistorta	−	+	0	0	0	0

while density reflected mostly the management regime.

Small scale dynamics of individual species

We observed two main types of species behaviour (Fig. 4): gradually decreasing correlation of species in a cell with increasing time lag and abrupt decrease of the correlation after a short time (usually one year) which then remained stable over a longer time. *Nardus stricta* and *Festuca rubra* show the former pattern, whereas *Polygonum bistorta* and *Deschampsia flexuosa* show the latter. *Anthoxanthum alpinum* differs between plots: in two plots (3 and 4, one manured and the other unmanured) it had the first pattern type and in the remaining plots the latter.

Interspecific associations

In spite of the superficial homogeneity of the grassland and of the similarity of the floristic (qualitative and quantitative) composition of the plots, the interspecific associations strongly differed among plots (Fig. 5). Interspecific associations were strongly dependent on the cell size. At the 1 × 1 cell level, only negative associations occur, and except for plot 2, these always contain *Nardus stricta* as one member of the pair. At larger cell sizes (2 × 2 and 3 × 3 cells), positive associations begin to prevail. Few species pairs show consistent behaviour in the whole set. For example, *Nardus stricta* is negatively associated with *Deschampsia flexuosa* in two plots at the 1 × 1 cell size, although the grain of this relationship differs between these plots: in plot 1, this negative association remains also at cell size 2 × 2, whereas in plot 3 it disappears at this cell size. Consistent behaviour is displayed e.g. by the pair *Nardus stricta–Anthoxanthum alpinum*, which shows positive association at the 2 × 2 size in all the plots. Incidentally, this is the only positive relation involving *Nardus stricta*. Some other species associations, though quite common, were displayed at different cell sizes (e.g. negative association *Festuca rubra–Nardus stricta*, which occurs at 1 × 1 size in plots 1 and 4, and at 2 × 2 size in plot 3). However, most of the rest of the species associations were limited to one plot only, though they might be very strong (e.g. *Festuca–Anthoxanthum* in plot 3/2 × 2). Species associations did not show any pronounced trends in time.

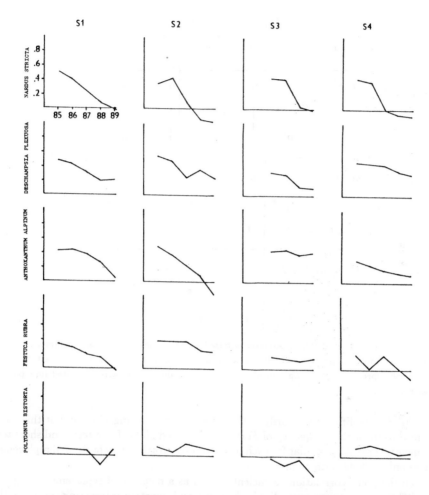

Fig. 4. Spatial change of component species. Abscissa expresses time interval since 1984, ordinate
the value of Spearman's correlation coefficient between the densities of the species at each
grid cell over this time interval. Plots S1 and S3 were manured in 1985.

Species replacements

Species replacements based on the 'a' cell of the 2 × 2 table differ rather a lot among
the four plots (Fig. 6). No species pair showed replacement in the same direction consistently in all four plots; only two species pairs (*Deschampsia-Polygonum, Deschampsia-
Anthoxanthum*) occurred in all plots, but the replacement direction varied. Fertilized and

179

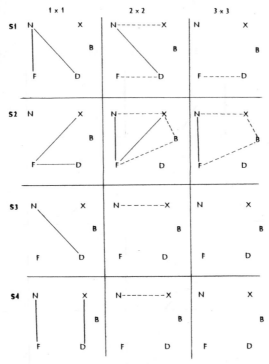

Fig. 5. Summary of interspecific associations based on the point correlation coefficient for each plot (lines indicate pairs in which the association had a value 0.15 for at least 3 times). Columns represent increasing cell sizes. Full-line indicates negative, dashed-line positive associations.

non-fertilized plots differed primarily in the interactions involving *Festuca*, but this has to be ascribed to the higher frequency of *Festuca* in fertilized plots. Long term and short term replacements were rather similar for each plot; however, *Nardus stricta* showed replacement processes only in long term comparisons.

When the point correlation coefficient is used as a measure of replacement (Fig. 7.), a quite different picture emerges. *Nardus* becomes an important interacting species within the community; most commonly it interacts with *Deschampsia* (though the direction varies among plots) both in long term and short term comparisons. In the long term comparisons, *Deschampsia* is replacing *Festuca* three times (out of four). In contrast, *Anthoxanthum* and *Polygonum* do not show many interactions either in short term and long term comparisons. In long term comparisons, *Nardus* is replacing *Festuca* in the manured plots; unfortunately, this relation cannot be compared to the unmanured plots because the number of sample cells is too low there.

In addition to the between plot variability, the direction of replacements in one plot varied also in time. The stability of the replacement processes is shown in Fig. 8. Impor-

180

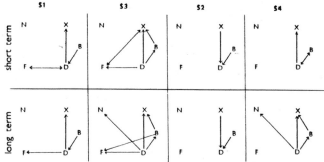

Fig. 6. Replacement processes defined by the 'a' cell in the four plots. An arrow indicates the particular replacement process running in the plot at least three times (out of five observation periods). The arrow is pointing in the direction of the replacing species. Upper row: short term replacement (over 1 year), lower row: long term replacement (over 5 years). Species abbreviations: B—*Polygonum bistorta*, D—*Deschampsia flexuosa*, F—*Festuca rubra*, N—*Nardus stricta*, X—*Anthoxanthum alpinum*. Plots S1 and S3 were manured.

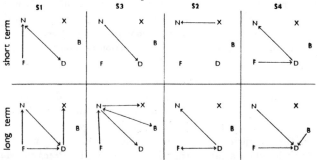

Fig. 7. Replacement processes defined by the point correlation coefficient in the four plots. For further explanation, see Fig. 6.

tantly, directional replacements occur (defined both by the 'a' cell and the point correlation coefficient) only in the manured plots.

Discussion

Species replacements based on the 'a' cell of the 2 × 2 table express absolute magnitudes of the replacement processes running in each plot, i.e the processes most important for plot dynamics. However, since they are dependent on species frequencies, they can be used for comparison of species between plots only if the species being compared do not differ in frequency (which condition is satisfied for most of the species in our plots). The point correlation coefficient values express replacement values standardized by frequencies. The large values thus indicate processes not necessarily most decisive at a given plot, but important from the point of view of the dynamics of the given species. The combination of both measures gives thus an extra depth of insight into the behaviour of the plants.

In this respect the difference in behaviour of *Nardus* between the two sets of replacement webs is very clear: *Nardus* is a rather slowly growing species forming compact tussocks

181

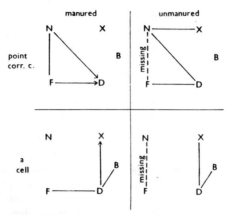

Fig. 8. Stability of the short term replacement processes based on the 'a' value (upper row) and on the point correlation coefficient (lower row) in fertilized and non-fertilized plots. A line indicates that the particular replacement process occurred at least five times (out of nine or ten observations for manured and unmanured plots respectively); an arrow is present wherever the number of replacements in one direction is lower than 1/2 of those in the other direction.

(phalanx strategy), as is also demonstrated by its fine scale dynamics. In the short time perspective, it does not play a great role in the community because of its slow growth. However, its interactions are highly specific as shown by the replacements measured by the point correlation coefficient. The effect of these interactions will become apparent at the community scale when a larger time interval is used. *Deschampsia* is, in a sense, a counterpart of *Nardus*: because of its high frequency and rapid change (see also its rapid qualitative decay curve in the small scale dynamics), it is central in the 'a' cell interactions; its effects are smaller when it is standardized or extended to the longer time scale. The same can also be said about *Anthoxanthum*.

Replacements between particular species in individual (unmanured) plots are not directional and thus do not lead to successional changes; as expected, manuring brings some directionality into the plots. Here the replacement processes show higher yearly rates, comparable with changes occurring in nonmanured plots over several years. The species having the highest positive response to manuring (*Festuca* and *Anthoxanthum*) increase their proportion; their absolute increase is hard to evaluate because of the large year-to-year variation in total biomass. Both these species differ in their response type: whereas *Anthoxanthum* decreases later (3–4 years) after the initial increase, the increase of *Festuca* is more stable.

The most striking thing in the whole observation is the very large variation of the species replacements. This variation has several components:

1. both symmetrical replacement processes (A replacing B, B replacing A) running simultaneously at one plot in one observation period;

182

2. asymmetric replacement in one plot alternating during the observation period, not necessarily simultaneously in all plots;

3. variability between plots.

Variation both in time and space is characteristic of grassland communities (see e.g. Rabotnov 1984, Mirkin 1985, van den Bergh 1979), often described as fluctuations of species dominance (Grubb *et al.* 1982, Rabotnov 1984, Rychnovská and Jakrlová 1990). Such variation is most probably due to just this variation of interspecific interactions. Since the importance of the first two components for maintenance of the diversity is obvious, both spatiotemporal variation and species richness of grasslands might be closely connected. This would corroborate current views on the role of the variability of external conditions in determining species diversity in communities (Levins 1979, Sebens 1987, Chesson 1988, Ebenhöh 1988, Grover 1988, Fowler 1988). The large interplot variability (third component) shows that also at a larger scale, equilibrium in grasslands is a result of the dynamics of highly diverse, non-equilibrium patches (cf. Mirkin 1985).

Processes operating within the community can be expected to be closely connected to the spatial pattern of the community (Watt 1947, During and van Tooren 1988). Influence of the spatial pattern on competition is sometimes studied (e.g. Schmid and Harper 1985, Pacala 1986); however, as the pattern is dependent on the growth of individuals, pattern influences competition only in slowly growing phalanx species (Schmid and Harper 1985). In a similar vein, replacement processes may be easily imagined to be responsible for the observed spatial pattern (e.g. interspecific associations), especially at the lowest spatial scale. In this respect the variability of the interspecific associations, which does not parallel the variability in replacements, is very striking. Thus the four studied plots, in spite of their superficial similarity, perhaps differ in some important respect not analysed by the present study. This is corroborated by the facts that (1) most of the species differ in their spatial autocorrelation patterns among the plots (unpubl. mater.), and (2) some species (*Anthoxanthum*, partly *Festuca*) differ also in their spatial stability curves.

Most probably, each of the three components of the replacement variability mentioned above is due to a different cause: whereas some sort of population process (e.g. spatial movement, ageing of the tussocks) or microenvironmental heterogeneity may be invoked in the first case, climatic change can be a tentative cause of the second and environmental heterogeneity of the third. However, these processes would only act as simple cause and effect if our grassland systems were otherwise at equilibrium. If the response time of the grass sward is sufficiently long, it may produce very complicated and possibly chaotic behaviour (Schaffer 1985), which, in turn, will make the history of the particular sward important. Indeed, the great differences in population structure of some species would indicate this. In such a case even the same external regime will not necessarily make the plots converge.

Acknowledgements. We are very grateful to the owners of the grassland, family Václav and Marie Luczkovi, for the possibility of working there and for their great hospitality during our stays in the field. We are also much indebted to the the former head of the geobotany department, Dr R. Neuhäusl, who enabled us to start this study. Numerous colleagues, namely Eduard Brabec, Hana Kotrbová, Miroslav Škoda, Denisa Blažková, Jiřina Květová, Sylvie Kettnerová, Kateřina Macurová, Anna Krahulcová, Alena Herbenová, and

Aleš Hájek, helped us with the data collection in the field.

References

Chesson P. L. (1988): Environmental variation and the coexistence of species.—In: Diamond J.A. and Case T.J. [eds.], Community ecology, Harper & Row, New York, pp. 240–256.

During H.J. and van Tooren B. (1988): Pattern and dynamics in the bryophyte layer of a chalk grassland.—In: During H.J., Werger M.J.A. and Willems J.H. [eds.], Diversity and pattern in plant communities, SPB Academic Publishing, The Hague, pp. 195–208.

Ebenhöh W. (1988): Coexistence of an unlimited number of algal species in a model system.— Theoret. Popul. Biol. 34: 130–144.

Fowler N. (1988): The effects of environmental heterogeneity in space and time on the regulation of populations and communities.—In: Davy A.J., Hutchings M.J. and Watkinson A.R. [eds.], Plant population ecology, Blackwell Sci. Publ., Oxford, pp. 249–269.

Grover J.P. (1988): Dynamics of competition in a variable environment: experiments with two diatom species.—Ecology 69: 408–417.

Grubb P.J., Kelly D., and Mitchley J. (1982): The control of relative abundance in communities of herbaceous plants.—In: Newman E. I. [ed.], The plant community as a working mechanism. British Ecol. Soc. Spec. Publ. 1, Blackwell, Oxford, pp. 79–98.

Hawksworth D.L. and Chater A.O. (1979): Dynamics and equilibrium in a saxicolous lichen mosaic.—Lichenologist 11: 75–80.

Kershaw K.A. (1973): Quantitative and dynamic plant ecology.—2nd ed., Arnold, London.

Krahulec F. (1990): *Nardo-Agrostion* communities in the Krkonoše and West Carpathians Mts.— Folia Geobot. Phytotax. 25: [in press].

Levins R. (1979): Coexistence in the variable environment.—Am. Nat. 114: 765–783.

Mirkin B.M. (1985): Teoreticheskie osnovy sovremennoy fytocenologii [Theoretical fundamentals of the modern plant ecology).—Nauka, Moskva.

Pacala S.W. (1986): Neighborhood models of plant population dynamics. 2. Multi-species models of annuals.—Theoret. Popul. Biol. 29:262–292.

Rabotnov T.A. (1984): Lugovedenie [Grassland science].—Izd. Mosk. Univ.

Rychnovská M. and Jakrlová J. (1990): Alternating dominance: a manifestation of ecological strategy in grassland stands.—In: Krahulec F., Agnew A. D.Q., Agnew S. and Willems J.H. [eds.], Spatial processes in plant communities, Academia (this volume), pp. 185–189.

Schaffer W.M. (1985): Order and chaos in ecological systems.—Ecology 66:93–106.

Schmid B. and Harper J.L. (1985): Clonal growth in grassland perennials. I. Density and pattern dependent competition between plants of different growth forms.—J. Ecol. 73: 793–808.

Sebens K.P. (1987): Competition for space: effects of disturbance and indeterminate competition success.—Theoret. Popul. Biol. 32:430–441.

van den Bergh J.P. (1979): Changes in the composition of mixed populations of grassland species.—In: Werger M.J.A. [ed.], The study of vegetation, Dr. W. Junk Publ., The Hague, pp.57–80.

van Rijnberk H. and During H. (1990): Bryophyte pattern and dynamics in Dutch chalk grasslands.—In: Krahulec F., Agnew A.D.Q., Agnew S. and Willems J.H. (eds.), Spatial processes in plant communities, Academia (this volume), pp. 161–172.

Watt A.S. (1947): Pattern and process in the plant community.—J. Ecol. 35:1–22.

Krahulec F., Agnew A.D.Q., Agnew S., Willems J.H. [eds.]: *Spatial processes in plant communities.* pp. 185–189.

Alternating dominance: a manifestation of ecological strategy in a seminatural Nardus stricta grassland

Milena Rychnovská[1] and Jana Jakrlová[2]

1) *Institute of Experimental Biology and Ecology of the Slovak Academy of Sciences (*author's present address: *Drobného 44, 602 00 Brno, Czechoslovakia)*
2) *Institute of Systematical and Ecological Biology, Czechoslovak Academy of Sciences, 662 61 Brno, Stará 18, Czechoslovakia*

Keywords: *Nardus stricta, Sanguisorba officinalis,* Production phenology, *Polygalo-Nardetum,* Transpiration rates.

Abstract. Seasonal variation in abundance of codominants has been observed in a polyspecies seminatural grassland dominated by *Nardus stricta* and *Sanguisorba officinalis.* The ecological strategies of these plants are different and their ecological niches are complementary. An ecophysiological explanation of this phenomenon is offered. Observation of the stand in consecutive years verified the idea that alternating dominance can be considered as one of the homeostatic mechanisms leading to the stability of this grassland.

Introduction

Seminatural and natural grasslands show a considerable stability which is based on many homeostatic mechanisms. We tried to analyse some of them and elucidate the remarkable ability of grasslands to cope with a large amplitude of abiotic factors in their habitat, e.g. with surplus, fluctuation or lack of soil water in their rhizosphere, with surplus or deficit in nutrients, wide range of pH, broad amplitudes of temperature, with disturbance of aerial parts, regular export of biomass, etc. Some of these mechanisms were evaluated in previous work (Rychnovská 1979, 1988). The present contribution is focussed on the relations between codominants in a seminatural unmanaged grassland in the Žďárské vrchy hills.

Results and Discussion

Plant community, the codominants and their characteristics

A seminatural grassland classified as *Polygalo-Nardetum strictae* Preising 1950 (Balátová-Tuláčková 1980) was analyzed in the wider framework of the MAB Field Project No. 91 ("Kameničky Project") during 1974 to 1985. The plant cover of the experimental plot 100 × 100 m consisted of about 64 species of vascular plants (Zelená 1982). The stand was neither cut nor fertilized during the research period and for several years before. The

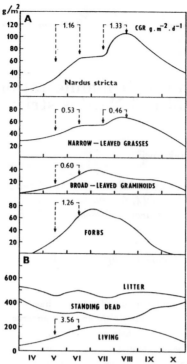

Figure 1. The yearly course of the aboveground biomass production of the *Polygalo-Nardetum strictae* at Kameničky. Means for 1976–1981. A — yearly variation in biomass of live shoots of the dominant and other morphological groups of vascular plants; B — yearly variation of total live biomass, standing dead material and litter. Numbers between arrows indicate the crop growth rate (CGR).

primary production was studied here for many years (Jakrlová 1979, 1980, Fiala 1979). Most of the area is co-dominated by *Nardus stricta, Festuca capillata, Sanguisorba officinalis, Potentilla erecta* and a few other species. Our attention and analyses were focussed on variation and adaptation of the two contrasting species, *Nardus stricta* and *Sanguisorba officinalis*, which can be considered as representative of the other codominants, narrow-leaved grasses and broad-leaved forbs respectively.

Proportions of the plant populations in the stand

Primary production of aboveground biomass was estimated by cutting the foliage just above the soil surface in 10 replicates of plots 10 × 40 cm. The proportion of particular species or morphological groups was expressed in grams (dry mass) per m². The sampling took place in 7 to 15 days' intervals during the growing season. Fig. 1 demonstrates the seasonal variation in aboveground biomass which persists here throughout the whole year. Nevertheless, a distinct growth activity is notable from April to August, when the living shoot biomass of the stand reaches its maximum value; it then slowly decreases

186

Table 1. Proportion of the codominants in the grassland community *Polygalo-Nardetum strictae* in consecutive years. Proportion of codominants in percentage of aboveground living biomass, coefficient of variation V equals standard deviation as a percentage of the arithmethic mean. Harvesting took place in July and represents the maximum standing crop.

	1976	77	78	79	80	81	82	83	84	V
Living shoot biomass (g m^{-2})	93	256	195	362	256	230	265	349	183	34
Nardus stricta (%)	24	23	43	28	24	31	34	19	11	35
Sanguisorba officinalis (%)	29	26	12	30	22	23	10	27	33	34
proportion of both codominants (%)	53	49	55	58	46	54	44	46	44	10

until November, when snow often covers the ground. A part of the foliage remains green, forming a permanent pool of living shoots. The majority, however, turns into standing dead biomass and litter. When analyzing separately the growth curves of *N. stricta* and narrow-leaved grasses on the one hand and the curves of broad-leaved grasses and forbs, represented mainly by *S. officinalis* on the other, an evident complementarity can be observed.

Narrow-leaved grasses show a slow growth in leaf biomass from April to June, and hardly any growth in July when the differentiation of generative organs is taking place. The production of summer tillers by these plants and their intensive growth result in a second growth maximum in August. Forbs participate in the stand only negligibly in April. Their rapid growth in May and June, however, and their fast penetration into the canopy of narrow-leaved grasses and, eventually, the formation of a new overstory by them, marks the peak of forbs in early summer. Broad-leaved graminoids have similar dynamics, except for their participation in the permanent pool of living biomass in winter.

Such a complementarity reflects the ecological niches and production strategy of the leading species, exploiting evenly in this way three months of the peak growing season when the primary production is least limited by unfavourable factors. The data on crop growth rate for the main growth period in Fig. 1 show the growth dynamics of both dominant groups. The second growth period in *N. stricta* corresponds to the decline of the forbs, contributing to a steady shoot biomass and assimilating surface during the most favourable conditions.

Alternating dominance in consecutive years

The amount of aboveground living biomass corresponding to the maximum standing crop is variable and depends on the climatic conditions of the particular year and on the underground reserves of the sward accumulated in the preceding year (Rychnovská *et al.* 1985). The proportions of the two codominants, however, are more or less constant and fluctuate around 50% of the total biomass, as shown in Table 1, they seem to be complementary. It could be assumed that both codominants utilize the resources of their habitat similarly. Perhaps very delicate feedback mechanisms controlled by the climate

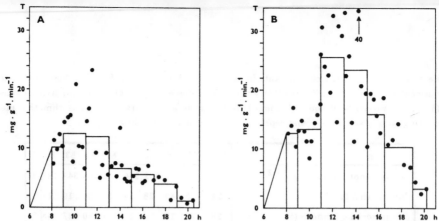

Figure 2. Daily variation of transpiration in the stand of *Polygalo-Nardetum strictae* at Kameničky, June 20, 1979. Expressed in mg min^{-1} per unit dry mass. A — transpiration of the entire stand; B — transpiration of *Sanguisorba officinalis*. Points correspond to individual measurements, columns to mean values. (After Rychnovská 1990.)

and by the adaptability of the plants determine the growth rate in a particular year and lead to the prevalence of one or the other codominant.

When extending the findings to the other constituents of the stand, a similar variation was observed. Among the 31 species whose shoot biomass was estimated quantitatively only nine species contributed to the standing crop in a measurable amount each year. The remaining 22 species contributed distinctly to the primary production in some years although they never disappeared from the stand completely. The ecosystem could be considered to consist of variable niches, which are promptly occupied by existing plant populations according to the seasonal variation of weather or different climatic or biotic stresses.

Physiological complementarity of codominants

A precise ecophysiological analysis of the species under study may help to explain the variation of their co-existence in the stand. The submitted analyses of the transpiration can be considered as the first approach. Analyses of photosynthesis of the sward and plant populations have been published elsewhere (Gloser 1990). The natural distribution of water between individual plant organs was studied first. It was found that the codominant forbs, i.e. *S. officinalis* and *Potentilla erecta* store more than 75% of their water reserves in underground organs, while up to 78% of the water content of fully turgid plants of *N. stricta* is stored in underground and nearground organs (roots, rhizomes, leaf sheaths) (Rychnovská 1988). It has been shown that these reserves stabilize water output and balance by providing water for transpiration and reducing the saturation deficit in the leaf blades during the noon hours (Rychnovská *et al.* 1980). From that point of view, the strategy of codominants, no matter which, seems to be similar. Accompanying species, e.g. *Succisa pratensis* and *Deschampsia cespitosa*, however, have more than 65% of their readily disposable water stored in the shoots. That may cause a higher sensitivity to contingent water stress and may contribute to their retreat in unfavourable years.

Another aspect is the transpiration. Fig. 2 shows a daily course of transpiration in the stand as compared with the water output of the dominant forb. The sward under

188

study consisted of 89% dry mass of graminoids and 11% of forbs, mainly *S. officinalis*.
Maximum transpiration rate of the stand measured at noon was about 12mg per g dry
mass per minute while the pertinent values for *S. officinalis* were about 30 mg g^{-1} min^{-1}.
A similar relation was also found in lowland meadows (Rychnovská 1976). Forbs show
usually higher transpiration rates than grasses in any type of grassland stand (Rychnovská
1990). This may lead to an idea of different ecological strategies of both components of
the sward. In our case, codominat *S. officinalis* displays the physiological features of a
competitor while narrow-leaved grasses should be considered more or less as stress-tolerant
plants (Grime 1979).

For a precise understanding of spatial processes in grassland swards, many more quantitative analyses of the variation of plant populations, their structures and physiological
functions should be performed besides monitoring the abiotic factors. This is the only way
we can begin to understand spatial and successional processes in plant communities.

References

Balátová-Tuláčková E. (1980): Übersicht der phytozönologischen Einheiten der Wiesen im
    Naturschutzgebiet Žďárské vrchy I.—Preslia 52: 311–331.
Fiala K. (1979): Estimation of annual increment of underground plant biomass in a grassland
    community (*Polygalo-Nardetum*).—Folia Geobot. Phytotax. 14: 1–10.
Gloser J. (1990): Canopy structure and production processes in two contrasting types of grass
    swards.—Ecology (ČSSR) 9: [In press].
Grime J.P. (1979): Plant strategies and vegetation processes.—Chichester, 222 pp.
Jakrlová J. (1979): Primary production of the Kameničky grassland—aboveground.—In: Rych-
    novská M. [ed.], Progress Report on MAB Project No 91: Function of Grasslands in Spring
    Region — Kameničky Project, Brno, pp. 70–86.
Jakrlová J. (1980): Changes of production of aboveground biomass and in plant assemblage of
    an uncut meadow stand in the course of several years.—Grassland Ecology. Proc. Meeting
    20–21 March 1980, Banská Bystrica, pp. 165–184 (in Czech).
Rychnovská M. (1976): Transpiration in wet meadows and some other types of grassland.—Folia
    Geobot. Phytotax. 11: 427–432.
Rychnovská M. (1979): Sources of stability in grassland ecosystems.—In: The 5th International
    Symposium on the Problems of Ecological Landscape Investigations, 19–23 November 1979,
    Vysoké Tatry. Bratislava, pp. 287–294.
Rychnovská M. (1988): The strategy of plants for maintaining their water balance.—In: Nakhut-
    srishvili G.Sh. [ed.], The Ecology of High Mountains. Tbilisi, pp. 222–229 [in Russian].
Rychnovská M. (1990): Water relations of individuals, plant populations and communities in a
    grassland ecosystem.—Ecology (ČSSR) 9: [In press].
Rychnovská M., Balátová E., Úlehlová B. et Pelikán J. (1985): Ecology of Meadows.—Academia,
    291 pp. [in Czech].
Rychnovská M., Čermák J. et Šmíd P. (1980): Water output in a stand of *Phragmites communis*
    Trin.—Acta Sci. Nat. Brno 14(2): 1–30.
Zelená V. (1982): Phytocoenological and floristic characteristic of plant communities studied in
    the surroundings of Kuklík and Kameničky villages.—Report on the Kameničky Project,
    Brno, No 32, pp. 15 [in Czech].

Krahulec F., Agnew A.D.Q., Agnew S., Willems J.H. [eds.]: *Spatial processes in plant communities.* pp. 191-199.

Resource pattern and community pattern in a dune slack grassland

Andrew D.Q. Agnew and Habiba Gitay

Department of Biological Sciences, University College of Wales, Aberystwyth, SY23 3DA, United Kingdom.

Keywords: Dune slack, Grain size, Phosphorus, Calcium

Abstract. Patterns in phytomass, plant phosphorus and organically based total soil phosphorus and calcium were examined in four stands of apparently uniform dune-slack grasslands of differing species richness. Previous analyses suggested that interactions between species were strongest in patches of 20 cm diameter and with increasing species diversity. We use the concept of environmental grain to discuss hypotheses of phosphorus and calcium distribution at this patch size and find that:

1. There is some evidence of phosphorus sinks developing around the more dominant species in less diverse communities, but not in the most species-rich system.

2. Phosphorus and calcium show similar distribution patterns, apparently unaffected by their differing mobilities, but calcium seems to be only negatively correlated with species phytomass at low diversities.

3. Although the more abundant species appear to become less correlated with the biological phosphorus pool as diversity increases, there is evidence that the underlying organically based soil pool of both phosphorus and calcium becomes more variable at the spatial dimension of species interactions.

4. The evidence suggests that at low diversities species are coarsening environmental grain, whereas at high diversities the environment is more patchy and occupied by species which exploit the patches in a fine grained way.

Introduction

In a previous paper (Gitay and Agnew 1989) we examined the nature of plant communities and whether species in assemblages interacted with each other to form patterns within a community. We found some evidence that there was true interaction but that this was only detectable in the most diverse communities. In our series of stands we tried to show that the highest diversities were not due to major heterogeneities in the resources available, using soil phosphorus, calcium and loss on ignition as environmental factors. However, heterogeneity in resource availability within a habitat may of course be at many scales, and here we attempt to refine our analyses to take into account the finer scales of resource heterogeneity which may be associated with species diversity and community structure.

191

There have been many suggestions that environmental heterogeneity is a principal cause of species diversity in communities (for instance Harper 1977) and indeed the classic experiments of Crombie (1947) with *Tribolium* beetles showed that uniform environments would select only one competitively successful species while heterogeneous environments allowed coexistence. Resources which are patchy in occurrence, or unpredictably presented in relation to an organism's search ability, have been called coarse grained, while fine grained resources are uniform and predictably available (MacArthur 1968). These concepts seem to us to be useful when dealing with the soil resource base for plant species with roughly similar dimensions, but we are aware that it is better in general to use the grain concept only in relation to an organism's exploration of its environment (Addison *et al.* 1987, Crawley 1983).

As an example of the use of grain concept we can re-interpret the reasonable assumption that species exist at a particular point in space because of the interplay between their resource and competitive environments. For instance analytical ordinations of small vegetation samples and accompanying environmental factors are often used to suggest associations between resources and species (Austin 1968, Barkham and Norris 1970, Causton 1988). In grain terminology these authors are assuming that species are able to track their requirements in a patchy environment by exploring in a fine grained way, so that the probability of each species finding its optimal micro-habitat is high.

But individual plants practise an economy of resources. To light and temperature they can equilibrate, but captured energy and nutrients are transferred, stored and eventually lost to adjacent litter, creating environmental heterogeneities which may coarsen the environmental grain particularly for other species. Chapin *et al.* (1979) have shown this very clearly in *Eriophorum vaginatum* which appears to create a sink of phosphorus in its dense tussock, depleting the surrounding area.

We examined a series of apparently homogeneous grasslands in sand dune slacks and found that competitive interactions may occur at a scale which was much larger than that of an individual or single sample (Gitay and Agnew 1989). We therefore suggest that there should be fine grained resource exploitation at this scale but coarser grain at the scale of individual plant size.This prediction is the basis for the present paper, where we attempt to relate the roughness of resource presence to species diversity and scale of community interaction.

Materials and Methods

We studied dune slack vegetation at Ynyslas National Nature Reserve, on the west coast of Wales, the same area as that mapped by Page *et al.* (1985); we have reported our methods more fully elsewhere (Gitay and Agnew 1989). Slack habitats have many advantages as sites in which to analyse community structure and function. They are constrained to a short season by a long period of high water table, and water movement should tend to redistribute soluble resources, thus keeping patterns of their distribution in a dynamic state but without residual influences of past events. Disturbances by wild fauna (principally rabbit grazing) although present are kept to a minimum by the short season. Although the dune system is comparatively young, slack soils are low in phosphorus and contain no free calcium carbonate (see below). The major species (Table 1) are perennial, most are hemicryptophytes, and most have the same strategy as "persistents" as defined by Loucks *et al.* (1985) for grassland species.

Four stands of differing species richness were chosen, and, in each of these, one transect of from 2460 to 4560 mm in length was taken, at almost the same date, and all plant material separated into species, in every small contiguous quadrat of $40mm \times 40$ mm. The morphology of certain of the plant species was such that separation into root and shoot material was impossible. For example, *Juncus articulatus* and *Hydrocotyle vulgaris* have superficial rhizomes and soil surface stolons respectively so that a distinction between below and above ground material was hard to make and led to no greater precision in the data. This quadrat size was the smallest we could use without incurring edge effect errors for individual plants, and most quadrats contained more than five species. Soil and root mass was taken to a depth of 80 mm. The rather low total standing crop listen in Table 1 demonstrates that these were low growing grasslands, with a canopy height of under 100 mm. The stands are here designated by the number of species abundant enough for use in statistical analyses in the 40×40 mm quadrats, i.e. 9, 12, 13 or 16, although the average number of species in each quadrat at 5.26, 7.63, 6.69 and 9.45 respectively shows that the two stands of medium diversity could be interchanged in the ordering.

Resources were assessed in the bulk ashed phytomass and soil. Only phosphorus was analysed in the phytomass while the soil was assayed for calcium as well. Differences in soil bulk recovered from the quadrats meant that spatial assessment of these resources was difficult. We assume that as far as the ecosystem is concerned most of the resource base of a quadrat should be in its organic fraction, and therefore use the amount of the two elements, calculated as a proportion of the percentage loss on ignition of the homogenised soil from each quadrat, as the basis for spatial analysis.

We used Hill's (1973) two-term local variance analysis (TTLVA) for all sizes of groupings of adjacent quadrats, from individual 40×40 mm up to half the total length of the transect. Kershaw (1957) has shown that this is a legitimate way of assessing pattern in individual species. We follow Kershaw in referring to these groups of quadrats as Block Sizes. Naturally soil and plant parameters can be treated in the same way in TTLVA.

Results

Table 2 displays the mean values encountered in each of the four resource parameters. Analysis of variance showed that all sites were significantly different for all parameters (Gitay 1987) but there is little evidence for a relationship between species richness and any of them. Our values for soil phosphorus are apparently rather higher than those reported for dune situations in western Europe (Pemadasa *et al.* 1974, Willis *et al.* 1959), although the various extraction methods used make comparisons difficult. In any case our levels are much lower than those from soils in disturbed landscapes (Lindsay and Vleck 1977, Mellinger and McNaughton 1975). Plant phosphorus seems, on the other hand, to be much lower in our dune slack plants than in other reported analyses (Ernst 1983, Kachi and Hirose 1983), so that the overall indication is that our dune slacks are indeed phosphorus deficient. Our calcium levels also seem to be rather lower than in many reports from dune slacks (Gibson 1984, Rozema *et al.* 1985, Willis *et al.* 1959) perhaps reflecting the low calcium carbonate content (3–4.5%) of the sand source from which the dunes have been built (Page *et al.* 1985).

193

Table 1. Species average phytomass in mg per quadrat of 40x40mm. Note that only those species which figured in analyses are cited, but the totals include all species.

	Species in stand:			
	9	12	13	16
Vascular plants				
Agrostis stolonifera	158.2	33.1	238.5	143.5
Carex arenaria	465.5	70.2		264.7
Carex flacca	699.3		545.5	422.5
Carex serotina		61.3		213.0
Dactylorhiza incarnata				14.4
Eleocharis palustris		18.9	379.5	
Eleocharis quinqueflora		27.2		
Eleocharis uniglumis	221.9			78.7
Equisetum variegatum		74.5	30.0	120.5
Galium palustre				2.6
Glaux maritima		217.9		66.9
Hydrocotyle vulgaris	106.4	20.5		6.2
Juncus articulatus	108.7	111.2	26.3	232.0
Juncus gerardii			34.4	
Leontodon taraxacoides			17.0	
Plantago coronopus			99.7	
Potentilla anserina		52.9	36.7	130.2
Prunella vulgaris				0.5
Ranunculus bulbosus	96.3			
Ranunculus flammula		36.0		136.0
Sagina maritima			0.7	
Trifolium fragiferum			1.7	
Bryophytes				
Amblystegium serpens	0.5			
Calliergon cuspidatum		102.1	47.6	87.5
Preissia quadrata	0.6			2.5
Tortula ruralis			15.1	
Total phytomass	1942.7	826.0	1478.8	1922.0

194

Table 2. Mean values for nutrient parameters used. These are mean concentrations as mg.kg^{-1} in each of the contiguous 40x40 mm quadrats.

Stand Species	Soil Calcium	Soil Phosphorus	Soil L.O.I.	Plant Phosphorus
9	20155	166	4.3	531
12	3654	173	3.2	762
13	8014	157	2.0	1045
16	15660	208	5.0	512

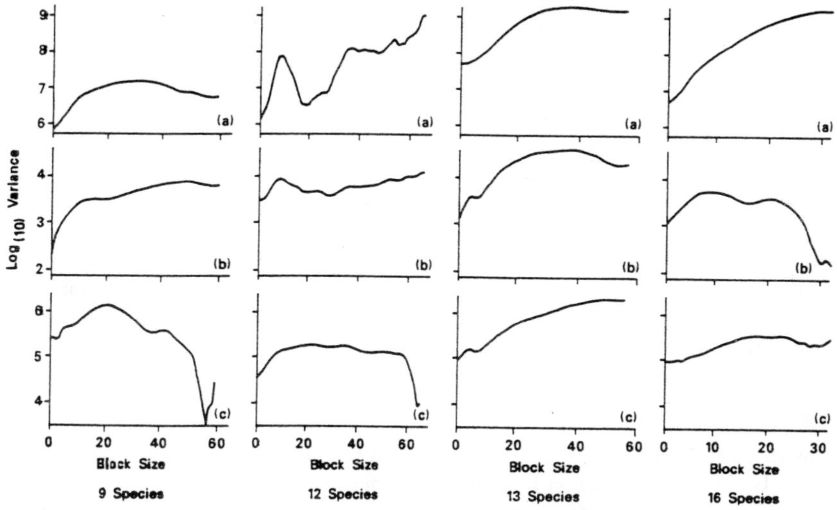

Fig. 1. Two term local variance levels for four stands of slack vegetation with species diversity as shown, (a) for calcium, (b) for soil phosphorus, (c) for total phosphorus.

Our previous paper (Gitay and Agnew 1989) gives our reasons for expecting major species' interaction and "community scale" to occur at around 20 cm diameter or Block Size 5 of the small quadrats. We therefore used this as the scale at which to examine the correlations between resources and species' phytomass within each stand. We did this for consistency, although we found no fundamental difference between patterns of resource correlations examined at Block Size 1, 5 or at the size at which each species showed a peak of variance. In Table 3 species with above average phytomass are cited separately in each stand to test the proposition that resources may pool around more successful species.

The relationship between the variances from TTLVA and Block Size is shown for the resources analysed in Figure 1. The tracks of variance show no clear general trend for the series of communities as a whole. There is no evidence of peaks or troughs at around Block Size 5, except in the 13 species stand, and this was our major reason for suspecting

195

Table 3. Summary of correlations $(r > 0.25, < -0.25)$ between species phytomass and resources analysed for all stands at Block Size 5. Major species are taken as those with phytomass above the stand median.

Stand species		Plant P		Soil P		Soil Ca		Total
		+	−	+	−	+	−	
9	major spp (5)	3	0	1	0	0	0	4
	minor spp (4)	0	0	0	0	0	0	0
12	major spp (6)	2	0	0	2	0	1	5
	minor spp (6)	0	0	0	0	0	2	2
13	major spp (6)	3	0	1	1	0	1	6
	minor spp (7)	1	0	0	1	0	2	4
16	major spp (8)	1	2	2	0	1	1	7
	minor spp (8)	3	0	2	3	1	0	9
	Total	13	2	6	7	2	7	37

that the mode of significant interspecific correlations at Block Size 5 was indeed due to species interactions at a scale of sociological pattern and not due to correlations with soil nutrients (Gitay and Agnew 1989). Soil phosphorus and calcium appear to have a similar structure within each stand.

Discussion

We know that resources can be patchy in uniform natural communities (e.g. Robertson *et al.* 1988). Is the plant community a force for coarsening the grain of resources or refining it, i.e. do the plants perform an operation on resources so that they are continually redistributed in a more patchy way or a more even way? Plants have differing resource requirements and efficiencies in acquiring those resources (Fohse *et al.* 1988, for crop plants, Jones 1975, specifically for our communities) as well as differing economies during their life histories (Gay *et al.* 1982). These effects should coarsen the scale of resources in a community so long as the individual plants making up the vegetation show a definite pattern of occurrence. Vitousek and Denslow (1986) have suggested a mechanism by which such a coarse scale could be maintained in forest stands through local enrichment around a fallen individual. On the other hand Atkinson (1973) has shown rapid movement and cycling of phosphorus through dune communities and this suggested that local concentrations will rapidly disperse through their surroundings. Indeed if plant species coarsen their environment we shall have difficulty in showing it, for the correspondence between scales of pattern in resources and vegetation would make causality impossible to assess.

We proposed in our earlier paper that our series of stands shows increasing organisation by guild formation and niche limitation as the number of species increases. Table 3 shows the differences between high and low species numbers in the nature of correlations

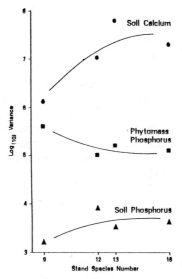

Fig. 2. Two term local variance at Block Size 5 (patches of 20 cm diameter) for each of the resources analysed and in each of the four dune slack stands of varying diversity.

between species and resources in our four stands. These must be taken only as indications of possible trends because of the low numbers of species involved and the absence of replicates, but nevertheless we believe they may identify hypotheses for testing on other model communities. Table 3 also shows that the proportion of significant correlations increases dramatically with increase in species number, which is another aspect of the trend in community organisation. There is also evidence for phosphorus pools developing in the 9, 12 and 13 species stands, because most correlations between total phytomass phosphorus and individual species' phytomass are positive and are formed in the species of greater importance in these three sites. A similar distribution of correlations might be obtained from a series of stands of decreasing dominance index (the proportion of total phytomass attributable to the two most abundant species in each stand (McNaughton 1968)), but the index is 60%, 40%, 62% and 36% respectively for the four stands in order of increasing diversity (Table 1): values which are inconclusive.

In any case the 16 species stand breaks the trend shown by the first three sites, for there are negative correlations and the minor species seem to be more associated with the two elements than are the major ones. Following our earlier argument then, there could be a process of coarsening grain in these communities until the number of species becomes so great as to introduce niche limitation through competition, at which point the major species are able to treat their resource environment in a fine grained way. This interpretation is very speculative of course, because the pooling of major species at low diversities seen in Table 3 is not echoed in the organic phosphorus and calcium pools and type I errors must occur when there are 50 possible correlations for each parameter.

If there is a coarsening effect on the soil environment at low diversities, and if the process ceases when the diversity is high enough, we would expect the most diverse communities to have the patchiest structure. Figure 2 shows the variance of the three parameters assessed over the four stands of differing species diversity. The differing scales of variance for the three nutrients are an artifact of the units involved in the analyses, but there are trends in variance which appear to follow changes in species richness. Thus Figure 2 appears to confirm the suppositions made from the correlation analysis: that the total phytomass pool of phosphorus becomes more evenly distributed while the soil resource pool of that element and calcium becomes more patchy as the number of species increases. The result is intuitively attractive, but does not resolve all the problems of the grain model because at high diversities we should see larger patches, within which a group of species could compete for a fine grained resource and Figure 1 contains no consistent evidence for patches of any size, or change in size. We subjectively judged that there was least disturbance in the 16 species site and most in the 13 species one, where bare soil was found. Like the analyses of mean calcium and phosphorus levels, these observations neither confirm nor negate our intuitive model of processes. Of great interest to us are the results of Glenn-Lewin and Ver Hoef (1988) who conceptualise their most species-rich sand prairie as consisting of small patches, within which species have a fine grained response to their environment, a conclusion very similar to ours.

This paper has aired a hypothesis about the organisation of communities with respect to resource pattern. But the imprecise nature of our analysis, the difficulty of proposing other model communities on which to test the hypotheses generated here, and the nature of spatial analysis which ignores time-based change must all make further progress very difficult.

References

Addison J.F., Aho J.M., Antolin M.F., Padilla D.K., Richardson J.S. and Soluk D.A. (1987): Ecological neighborhoods: scaling environmental patterns.—Oikos 49: 340–346.

Atkinson D. (1973): Observation on the phosphorus nutrition of two sand dune communities at Ross Links.—J. Ecol. 61: 117–133.

Austin M.P. (1968): An ordination study of a chalk grassland community.—J. Ecol. 56: 739–757.

Barkham J.P. and Norris J.M. (1970): Multivariate procedures in an investigation of vegetation and soil relations of two beech woodlands.—Ecology 51: 630–639.

Causton D.R. (1988): An introduction to vegetation analysis - principles, practice and interpretations. pp. 342. Unwin-Hyman, London.

Chapin F.S., III, Van Cleve K. and Chapin M.C. (1979): Soil temperature and nutrient cycling in the tussock growth form of *Eriophorum vaginatum*.—J. Ecol. 67: 169–189.

Crawley M.J. (1983): Herbivory.—Studies in Ecology 10.

Crombie A.C. (1947): Interspecific competition.—J. Animal Ecol. 16: 44–73.

Ernst W.H.O. (1983): Element nutrition of two contrasted dune annuals.—J. Ecol. 61: 99–105.

Fohse D., Claassen N. and Jungk A. (1988): Phosphorus efficiency of plants I. External and internal requirements and phosphorus uptake efficiency in different plant species.—Plant & Soil 110: 101–109.

Gay P.E., Grubb P.J. and Hudson H.J. (1982): Seasonal changes in the concentrations of nitrogen, phosphorus and potassium and in the density of mycorrhiza in biennial and matrix-forming species of closed chalkland turf.—J. Ecol. 70: 571–193.

Gibson D.J. (1984): Small scale pattern and soil heterogeneity in grassland.—Ph. D. thesis, University of Wales.

Gitay H. (1987): Plant community structure in dune slacks.—Ph. D. Thesis, University of Wales.

Gitay H. and Agnew A.D.Q. (1989): Plant community structure, connectance, niche limitation and species guilds within a dune slack grassland.—Vegetatio 83: 241–248.

Glenn-Lewin D.C. and Ver Hoef J.M. (1988): Scale, pattern analysis and species diversity in grasslands.—In: During H.J., Werger M.J.A. and Willems J.H. [eds.], Diversity and pattern in plant communities, pp. 115–129.

Harper J.L. (1977): Population biology of plants.—Academic Press, London.

Hill M.O. (1973): The intensity of spatial pattern in plant communities.—J. Ecol. 61: 225–235.

Jones R. (1975): Comparative studies of plant growth and distribution in relation to waterlogging VIII. The uptake of phosphorus by dune slack plants.—J. Ecol. 63: 109–116.

Kachi N. and Hirose T. (1983): Limiting nutrients for plant growth in coastal sand dune soils.—J. Ecol. 71: 937–944.

Kershaw K.A. (1957): The use of cover and frequency in the detection of pattern in plant communities.—Ecology 38: 291–299.

Lindsay W.L. and Vleck P.L.G. (1977): Phosphate minerals.—In: Dixon J.B. and Weed S.B. [eds.], Minerals in soil environments, pp. 639–672.

Loucks O.L., Plumb-Mentjes M.L. and Rogers D. (1985): Gap processes and large scale disturbances in sand prairies.—In: Pickett S.T.A. and White P.S. [eds.], The ecology of natural disturbance and patch dynamics, pp. 72–85.

MacArthur R. (1968): The theory of the niche. In: Lewontin R.C. [ed.], Population biology and evolution, pp. 159–176.

McNaughton S.J. (1968): Structure and function in Californian grasslands.—Ecology 49: 962–972.

Mellinger M.V. and McNaughton S.J. (1975): Structure and function of successional vascular plant communities in central New York.—Ecol. Mon. 45: 161–182.

Page R.R., da Vinha S.G. and Agnew A.D.Q. (1985): The reaction of some sand dune species to experimentally imposed environmental change: a reductionist approach to stability.—Vegetatio 61: 105–114.

Pemadasa M.A., Greig-Smith P. and Lovell P.H. (1974): A quantitative description of the distribution of annuals in the dune system at Aberffraw, Anglesey.—J. Ecol. 62: 379–402.

Robertson G.P., Huston M.A., Evans F.C. and Tiedje J.A. (1988): Spatial variability in a successional plant community: patterns of nitrogen availability.—Ecology 69: 1517–1524.

Rozema J., Laan P., Brockman R., Ernst W.H.O. and Appelo C.A.J. (1985): On the line transition and decalcification in the coastal dunes of north Holland and the island of Schiermonnikoog.—Acta Bot. Neerl. 34: 395–411.

Vitousek P.M. and Denslow J.S. (1986): Nitrogen and phosphorus availability in treefall gap of a lowland tropical forest.—J. Ecol. 74: 1167–1178.

Willis A.J., Folkes B.F., Hope-Simpson J.F. and Yemm E.W. (1959): Braunton burrows: the dune system and its vegetation, I.— J. Ecol. 47: 1–24.

Krahulec F., Agnew A.D.Q., Agnew S. & Willems J.H. [eds.]: *Spatial processes in plant communities.* pp. 201–213.

Changes in the spatial pattern of vegetation structure and of soil properties in early old-field succession

Ewa Symonides and Urszula Wierzchowska

Institute of Botany, Warsaw University, Al. Ujazdowskie 4, 00-478 Warsaw, Poland

Keywords: Old-field succession, Species richness, Plant biomass, Point diversity, Soil properties, Environmental heterogeneity

Abstract. Analyses of spatial distribution of species number, their above-ground biomass and point diversity (sensu Whittaker 1977), as well as of some soil parameters: pH, organic carbon and metal cation content, hydrolytical acidity and soil saturation with metal cations were conducted in four adjacent old fields situated within a single habitat. Plant and soil samples were collected within a grid of 100 shifted squares, 1 m² each. It was found that during early phases of secondary succession crucial changes in spatial vegetation structure and soil properties took place. The soil heterogeneity increased both horizontally and vertically but most vegetation and soil characters were not closely correlated.

Introduction

Our present knowledge of secondary succession comes above all from "old-field succession", as does much modern theory starting with Egler's (1954) classic paper on the importance of initial floristic composition (cf. van der Maarel 1988). The large number of studies on such secondary succession have proved, however, that general conclusions are not easily drawn. For instance, well-known, rapid old-field succession from short-living therophytes to long-living phanerophytes of clearings and woodlands takes place on loamy, but not on sandy soils (Schmidt 1988). Also that age in which an old field plant community exhibits the highest species richness and diversity may vary according to soil properties (cf. Houssard *et al.* 1980, Symonides 1985a). Moreover, all three mechanisms included in the approach of Connell and Slatyer (1977) appear to operate simultaneously in various parts of a single old field (Bornkamm 1988).

The above examples indicate that: (1) edaphic factors may, to a large extent, affect the rate at which the old-field vegetation attains its terminal stage, and (2) the course of succession may differ in various parts of a single old field due to environmental heterogeneity. However, no empirical proof of the latter has yet been found.

The present paper is part of wide-ranging studies on the process of old-field succession and spontaneous regeneration of subcontinental pine forest (*Peucedano-Pinetum* Mat. (1962) 1973 — Symonides 1985a, 1985b, 1986, Symonides and Borowiecka 1985). It describes a preliminary analysis of the direction and rate of change in spatial pattern of

a

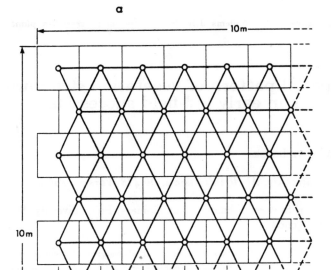

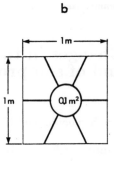

Fig. 1. Sampling scheme. (a)—grid of shifted squares, (b)—position of a circle within a square.

vegetation and spatial heterogeneity of a biotope during old-field succession. The studies aim to verify the following hypotheses: (1) spatial variation of vegetation and soil characters increases with old-field age, and (2) changes in spatial vegetation pattern are correlated with changes in soil properties.

Study site

The field study has been carried out at the southwestern edge of the Bialowieża Primeval Forest (Poland) within its historical XVIth century boundaries. In the late XVIth century some forests were cut down and replaced by arable fields. In 1917 the reverse process of forest re-invasion into abandoned fields commenced. Over the next 70 years a climax forest community (*Peucedano-Pinetum*) developed in a series of eight phases (Faliński 1980).

Four fields, 8, 12, 18 and 27 years old, were chosen for the analysis of soil and vegetation. They represented the 3rd, 4th, 5th and 6th successional phase, respectively. The study areas were located close to each other, in the vicinity of small, scattered pine forests. The nearest settlements are situated ca. 2.5 km away. Thus spontaneous succession in abandoned fields has been disturbed by neither direct (nor indirect) human impact.

Material and Methods

Samples of plants and soil were collected in a high season (July) from a grid of 100 shifted squares, 1 m² each (Fig. 1). The centre of each square was also the centre of 0.1 m² circle, within which above-ground parts of all plants were cut off, and 15 soil samples were taken from horizon I (5–10 cm, top layer) and 15 from horizon II (20–25 cm, bottom layer), using sampling stick.

The plants were sorted out into species, dried at 105°C for 48 hours and weighted with an accuracy of 0.001 g. Only fragile *Cladonia* thalli were not separated. For each circle the number of species, their total above-ground biomass and that of each species were estimated. On these grounds point diversity (sensu Whittaker 1977) was calculated, followed by the analysis of the spatial pattern of species number, their biomass and point diversity in all the old fields studied.

From soil samples (each was a mixture composed of 15 small samples) the following parameters were assessed: pH_{KCl} (potentiometrically), organic carbon, C_{org} (by Thiurin method), contents of four metal cations: Ca^{++}, Mg^{++}, K^+ and Na^+ (by spectroscopy), and hydrolytical acidity, H_h (by Kappen method, modified by Adrian). Next total exchangeable bases (S) and soil saturation with metal cations (V_S) according to the formula: $V_S = 100S/(S + H_h)$ were calculated. For all the parameters the ratio of their values in the top layer (I) to those in the bottom layer (II) was considered to be a differentiation index of upper soil horizons. The spatial pattern of this index was further analysed for all old fields.

For each parameter mean, standard deviation and the coefficient of variation, as well as correlation coefficients between the characters of vegetation structure and soil parameters were calculated. All calculations concerning pH were based on specific acidity (Wherry 1920).

Spatial pattern of vegetation and soil properties was assessed on the basis isarithmic charts obtained through mathematical interpolation (Kocimowski and Kwiatek 1976). This paper uses only the most representative fragments of maps reflecting both the scale and intensity of spatial structure of selected parameters.

Results

The studies show that during old-field succession not only do mean values of individual characters of vegetation structure and of soil chemical parameters change, but also, and above all, their spatial patterns change. Moreover, the nature, intensity and rate of such changes vary. For vegetation, considerable differences arise in both mean values of individual properties and their spatial pattern. On the other hand, the changes in soil parameters occur mainly in their spatial pattern, while means do not differ significantly.

Mean species number per 0.1 m² changes significantly in the course of the analysed succession. In the period under study it decreases almost threefold, although differences between the 18 and 27 years old fields are already negligible. The coefficient of variation increases as well. The rate of decrease in the value of species number and of increase in its coefficient of variation become reduced with the age of the old field (Table 1).

Also spatial pattern of species number undergoes considerable changes. The areas with the highest values, dominating in the youngest old field, diminish, while those with the lowest values (appear and) increase. Hence, the spatial differentiation of this parameter at first increases, then decreases. The mosaic of patches with highly diversified species number is most marked in the 8th succession year (Fig. 2).

Mean biomass of above-ground plant parts per 0.1 m² undergoes even bigger changes, but no obvious trend has been observed. In old fields, from the youngest to the oldest, at first it decreases (rapidly), minimum in the fifth phase, and then, (also rapidly) increases

Table 1. The comparison of means, standard deviations (S.D.), and coefficients of variation (V) of some vegetation characteristics between old fields differing in age.

| characters | old-field age (years) | | | | | | | | | | | |
| | 8 | | | 12 | | | 18 | | | 27 | | |
	x̄	S.D.	V (%)	x̄	S.D.	V (%)	x̄	S.D.	V (%)	x̄	S.D.	V (%)
Species number	9.5	2.53	26.7	7.1	2.67	37.6	3.83	1.48	38.6	3.78	1.48	39.2
Biomass (g.0.1m^{-2})	25.4	17.5	68.8	19.7	15.6	79.3	9.9	7.5	75.4	29.7	28.8	96.8
Point diversity (H')	0.49	0.15	30.6	0.42	0.15	35.7	0.32	0.16	50.0	0.31	0.17	54.8

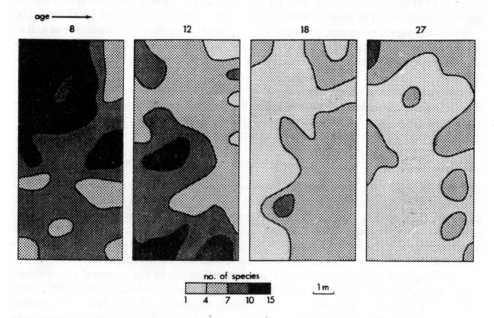

Fig. 2. Changes in spatial pattern of species number per unit area during succession.

(Table 1). Irrespective of the successional phase the biomass has highly clumped spatial distribution, however, the scale of the structure, and above all its intensity vary. The contrast between the patches with different biomass values diminishes between the 8th and 18th succession year, but later grows rapidly (Fig. 3).

The changes in point diversity are directional, but not so big as in the above mentioned two properties. Gradual reduction of mean point diversity is accompanied by a considerable increase in the value of its coefficient of variation (Table 1).

Spatial pattern of point diversity undergoes directional, and also fairly big changes. The whole H' range was found only for the youngest old field. In the course of the succession

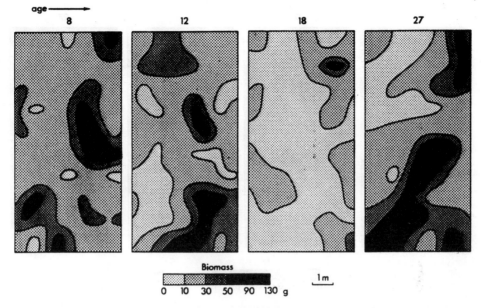

Fig. 3. Changes in spatial pattern of above-ground biomass during succession.

the patches with high diversity disappear, while those with its lowest values become more frequent. The rate of change is faster in early succession and later on slows down.

All analysed soil parameters are less variable in time than the vegetation characters. Compared with mean values bigger although not always directional changes concern their spatial pattern, in both analysed (top and bottom) layers. Mean values, except pH, do not differ significantly between the all old fields considered (Table 2).

The case of pH well illustrates the nature of changes in the habitat heterogeneity during old-field succession. Its mean gradually decreases in the top soil layer, showing its growing acididy. Spatial differentiation of this parameter is only slight in the youngest old field. Later however, it becomes more complex and results in a mosaic of patches with most different pH values in the 18 year old-field, then it considerably decreases. Mean pH is slightly higher in the bottom than in the top layer and in fact does not change with the age of an old field. Also its spatial differentiation becomes complex much later. Small patches with pH ranging from 3.75 to 3.85 appear as late as in the last phase, while in the top layer such mosaics represented in the youngest old field (Fig. 5a, b).

A much higher rate of pH changes in the top than in the bottom layer results in significant, directional changes in spatial pattern of pH_I/pH_{II} index. Upper horizons, almost homogeneous at first, become diversified and are transformed into a mosaic of small patches with highly different index values (Fig. 6).

Mean values of other soil parameters studied also vary negligibly in the course of the succession. The changes are either directional, e.g. a gradual decrease in metal cations contents (S) and soil saturation with metal cations (V_S), or in a form of fluctuations,

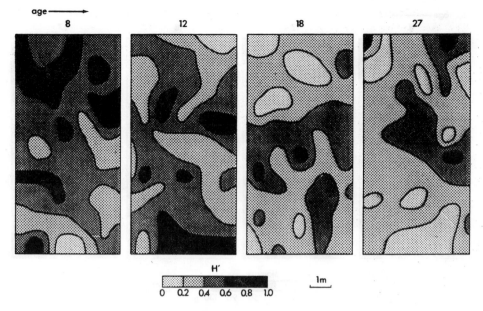

Fig. 4. Changes in spatial pattern of point diversity during succession.

like organic carbon content (C_{org}.) and hydrolytical acidity (H_h). Mean values of all parameters are significantly lower in the bottom than in the top soil layer (Table 2). Spatial patterns of differentiation of upper horizons during succession differ between properties (parameters).

In the youngest phase the values of C_I/C_{II} ratio are similar within the analysed area and range from 1 to 1.5. Hence the organic carbon content is the same or almost the same in both analysed soil layers. In further phases increasing or decreasing carbon content varies regarding its rate, intensity, and horizontal and vertical distribution within the old field. This causes growing differences between layers, and spatial mosaic of patches with various index values. Thus, in the oldest field only in a few sites does carbon content not depend upon the depth of the soil profile. Usually, its values are 2.5-3.0 times higher in the top than in the bottom layer (Fig. 7).

Similar but less obvious changes in the spatial differentiation of upper soil horizons concern the total metal cations. Although the range of S_I/S_{II} values is the same regardless of the successional phase. The proportion of patches with high metal content in the top layer increases, while that with similar content in both layers decreases gradually (Fig. 8).

Hydrolytical acidity is similar in the analysed soil layers, hence the values of $H_{h,I}/H_{h,II}$ index oscillate between 1 and 1.7 irrespective of the old field age. The variation in spatial pattern of the index is non-directional; in consecutive phases the patches with low or fairly high values dominate in turn (Fig. 9).

The last analysed parameter, soil saturation with metal cations, does not exhibit any significant changes in the spatial pattern of $V_{S,I}/V_{S,II}$ index. In all successional phases the

Table 2. The comparison of means (x), standard deviations (S.D.), and coeficients of variation (V) of some soil properties between old-fields differing in age (I—top layer (5–10cm), II—bottom layer (20–25cm)).

Property	Layer	old-field age (years) 8			12			18			27		
		x̄	S.D.	V(%)	x̄	S./D.	V (%)	x̄	S.D.	V (%)	x̄	S.D.	V (%)
pH(KCl)	I	4.0	-	15.9	3.9	-	15.5	3.8	-	23.1	3.7	-	24.7
	II	4.1	-	15.6	4.0	-	15.3	4.0	-	15.4	4.0	-	16.2
C(org)(%)	I	1.04	0.2	19.2	1.05	0.19	18.1	1.18	0.29	24.6	1.02	0.25	24.5
	II	0.77	0.15	19.5	0.67	0.11	16.4	0.80	0.13	16.3	0.60	0.05	8.3
S	I	0.44	0.30	68.1	0.34	0.14	41.2	0.32	0.30	93.7	0.24	0.17	70.9
(me/100g soil)	II	0.27	0.18	66.2	0.18	0.08	44.2	0.13	0.07	53.8	0.11	0.05	45.5
H(h)	II	8.86	0.66	7.5	8.81	0.85	9.7	9.16	1.13	12.3	8.25	0.84	10.2
(me/100g soil)	II	7.84	0.66	8.4	7.33	0.80	10.9	8.59	0.86	10.0	6.56	0.58	8.9
V(S)	I	4.55	2.53	55.6	3.67	1.53	41.7	3.17	2.52	79.5	2.71	1.65	60.9
(%)	II	3.23	1.91	59.1	2.31	1.06	45.9	1.46	0.65	30.5	1.65	0.69	41.2

patches with low values dominate indicating a similar direction and rate of change in both soil layers. Regardless of the field age there also occur patches, smaller or bigger, with V_S values much higher in the top soil layer (Fig. 10).

Discussion

The role of soil properties in determining the nature of plant communities is still very controversial. However, it has long been emphasized that the processes of succession and soil development are connected (cf. Moravec 1969). On the other hand, it is commonly known that on soils of a single type different communities can occur, and that the same vegetation units can occupy different soil types (cf. Neuhäusl and Neuhäuslová 1968). Also as yet it has not been determined which factors are responsible for the (inner) structure of a plant community. Some data indicate that the key role is played by a small-scale variation in soil properties through its impact on plant growth and competitive relationships within a plant community (cf. Hanawalt and Whittaker 1976, Tilman 1982, Goldberg 1987). However, according to other sources (Reader and Buck 1976, Tilman 1985, Bradshaw and Goldberg 1989) diversified light influx within a habitat, which plants compete for more intensely than they do for nutrients or water, can be of more importance.

Much published work, therefore, shows that soil is only one, and not even the most important, factor responsible for the species composition and spatial structure of plant communities. Such a conclusion can be drawn also on the basis of the present results. Although they show changes in spatial patterns of both vegetation structure and soil parameters during succession, their correlation has not been confirmed.

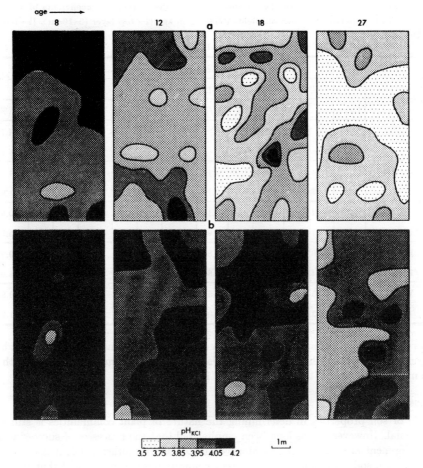

age ⟶

8   12   a   18   27

b

pH$_{KCl}$

3.5 3.75 3.85 3.95 4.05 4.2

1m

Fig.5. Changes in spatial pattern of pH in top (a) and bottom (b) soil layer during succession.

An increase in the homogeneity of species spatial pattern in the compared plant communities results undoubtedly from a decrease in species number, typical of a secondary succession (cf. Odum 1969, Shafi and Yarranton 1973, Bazzaz 1975). In these studies on old-field succession it has been considered to be a result of rapid recession of arable weeds and slow encroachment of forest species.

According to Quarterman (1957), Nicholson and Monk (1974), Houssard *et al.* (1980) a decrease in species number several years after the abandonment of cultivation is caused mainly by the progressive mineralization of the humus horizon and by soil depletion. The present results suggest that, as changes in chemistry of poor soils are negligible, competitive relationships in old-fields have more influence. In an undisturbed environment mosses, lichens and perennial plants grow luxuriantly so worsening the conditions for germination

208

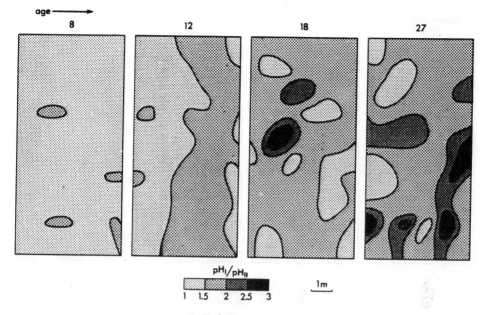

Fig. 6. Changes in spatial pattern of pH_I/pH_{II} ratio during succession.

and development of annuals and biennials. Thus, in the course of succession a gradual increase in the abundance and cover of the mosses, lichens and perennial plant species simultaneously with a decrease in participation of the short-lived species takes place (Gross and Werner 1982, Symonides 1986, Goldberg and Gross 1988, Bradshaw and Goldberg 1989).

Also soil properties and their spatial variation can hardly be considered a key agent determining the type of spatial biomass distribution found here. Spatial habitat differentiation, contrary to the opinion of some authors (e.g. Kershaw 1958, Greig-Smith 1961, Harper et al. 1965, Tilman 1985), plays only a minor role. It is indicated by a high similarity of patterns between the oldest and youngest old-field, and a lack of correlation between biomass pattern and any of the analysed soil parameters. A decrease in species number and transformation of floristic composition are also of some importance. Regardless of the old-field age 82–98% of total biomass is contributed by a few species present within the whole analysed succession period. Hence, it seems that biomass spatial pattern in the old fields under study results from biological and morphological properties of dominants (cf. Kwiatkowska 1972). It seems worth stressing that the relation between biomass, species number and soil properties may vary with the vegetation type and environment as a whole (cf. Vermeer and Berendse 1983).

Point diversity, fairly low irrespective of the old-field age, has been found to decrease in early succession, like alpha diversity (Nicholson and Monk 1974, Bazzaz 1975, Symonides 1985b). Again, such a result does not confirm the theoretical assumption that an increase in spatial habitat heterogeneity leads, especially on poor soils, to increased species di-

209

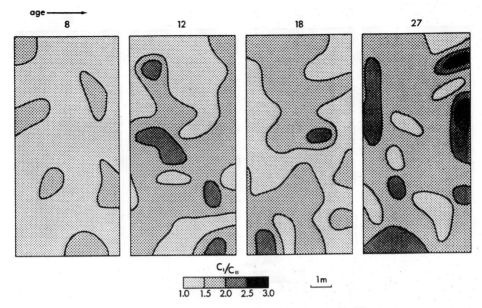

age ⟶
8   12   18   27

C_I/C_{II}

1.0 1.5 2.0 2.5 3.0

1m

Fig. 7. Changes in spatial pattern of organic carbon content ratio (C_I/C_{II}) during succession.

versity (cf. Tilman 1982). However, it seems possible that further differentiation on the environment in consecutive successional phases will be accompanied by a growth both in mean point diversity and its spatial variation to the level of climax pine community (cf. Kwiatkowska and Symonides 1986).

Summing-up, these studies on early old-field succession, have shown that the properties of poor soils change more slowly than does vegetation structure. Over a fairly wide range of soil (parameter) variation the vegetation is determined by such factors as growth habit and competitive ability of early colonizers.

Acknowledgement. We would like to thank Mrs K. Czerwińska and E. Cendrowska for soil analyses, and also Mrs A. Krzyżanowska-Mazur for translation of the text.

References

Bazzaz F. A. (1975): Plant species diversity in old-field successional ecosystems in southern Illinois.—Ecology 56: 485–488.

Bornkamm R. (1988): Mechanisms of succession on fallow lands.—Vegetatio 77: 95–101.

Bradshaw L. and Goldberg D. E. (1989): Resource levels in undisturbed vegetation and mole mounds in old fields.—Am. Midl. Nat. 121: 176–183.

Connell J. H. and Slatyer R. O. (1977): Mechanisms of succession in natural communities and their role in community stability and organization.—Am. Nat. 111: 1119–1144.

Egler F. E. (1954): Vegetation science concept. I. Initial floristic composition, a factor in old-field vegetation development.—Vegetatio 4: 412–417.

210

age ⟶

8 12 18 27

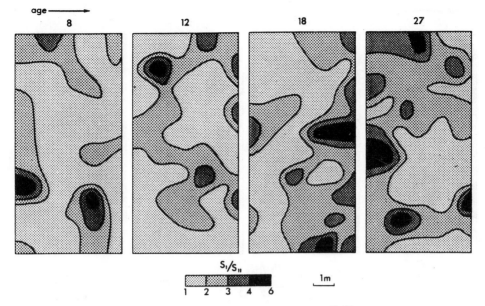

S_I/S_{II}

1 2 3 4 6

1m

Fig. 8. Changes in spatial pattern of metal cation content ratio (S_I/S_{II}) during succession.

Faliński J. B. (1980): Vegetation dynamics and sex structure of the populations of pioneer dioecious woody plants.—Vegetatio 43: 23–38.

Goldberg D. E. (1987): Seedling colonization of experimental gaps in two old–field communities.— Bull. Torr. Bot. Club 114: 139–148.

Goldberg D. E. and Gross K. L. (1988): Disturbance regime of midsuccessional old fields.— Ecology 69: 1677–1688.

Greig-Smith P. (1961): Data on pattern within plant communities. I. The analysis of pattern.—J. Ecol. 49: 695–702.

Gross K. L. and Werner P. A. (1982): Colonizing abilities of "biennial" plant species in relation to ground cover: implications for their distributions in a successional sere.—Ecology 63: 921–931.

Hanawalt R. B. and Whittaker R. H. (1976): Altitudinally coordinated patterns of soils and vegetation in the San Jacinto Mountains, California.—Soil Science 121: 114–124.

Harper J. L., Williams J. T. and Sagar G. R. (1965): The behaviour of seeds in soil. I. The heterogeneity of soil surface and its role in determining the establishment of plants from seeds.—J. Ecol. 53: 273–286.

Houssard C., Escarre J. and Romane F. (1980): Development of species diversity in some mediterranean plant communities.—In: van der Maarel E. [ed.], Succession, pp. 59–72. Junk, Hague.

Kershaw K. A. (1958): An investigation of the structure of a grassland community. I. The pattern of Agrostis tenuis.—J. Ecol. 46: 571–592.

Kocimowski K. and Kwiatek J. (1976): Wykresy i mapy statystyczne.—Główny Urzad Statysty-

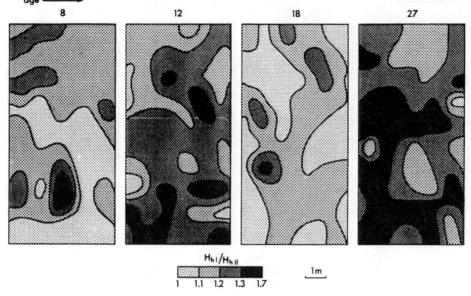

Fig. 9. Changes in spatial pattern of hydrolytical acidity ratio ($H_{h,I}/H_{h,II}$) during succession.

czny, Warszawa, 155 pp.

Kwiatkowska A. J. (1972): Analiza homogeniczności runa fitocenozy przy zastosowaniu nieparametrycznego testu serii.—Phytocoenosis 1: 37–77.

Kwiatkowska A. J. and Symonides E. (1986): Spatial distribution of species diversity indices and their correlation with plot size.—Vegetatio 68: 99–102.

Moravec J. (1969): Succession of plant communities and soil development.—Folia Geobot. Phytotax. 4: 133–164.

Neuhäusl R. and Neuhäuslová Z. (1968): Mesophile Waldgesellschaften in Südmähren.—Rozpr. Čs. Akad. Věd. cl. math.-nat. 78/11.

Nicholson S. A. and Monk C. D. (1974): Plant species diversity in old-field succession on the Georgia Piedmont.—Ecology 55: 1075–1085.

Odum E. P. (1969): The strategy of ecosystem development.—Science N. Y. 164: 262–270.

Quarterman E. (1957): Early plant succession on abandoned cropland in the Central Basin of Tennessee.—Ecology 38: 300–309.

Reader R. H. and Buck J. (1986): Topographic variation in the abundance of *Hieracium floribundum*, relative importance of differential seed dispersal, seedling establishment, plant survival and reproduction.—J. Ecol. 74: 815–822.

Schmidt W. (1988): An experimental study of old-field succession in relation to different environmental factors.—Vegetatio 77: 103–114.

Shafi M. J. and Yarranton G. A. (1973): Diversity, floristic richness and species evenness during a secondary (postfire) succession.—Ecology 54: 897–902.

age ⟶

8   12   18   27

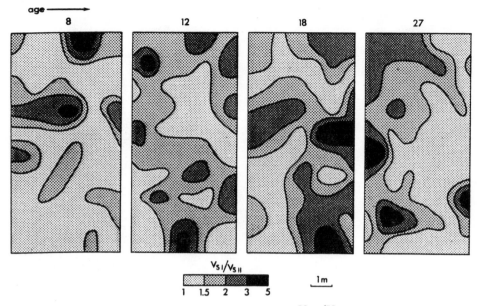

$V_{S,I}/V_{S,II}$

1 1.5 2 3 5   1 m

Fig. 10. Changes in spatial pattern of soil saturation ratio ($V_{S,I}/V_{S,II}$) during succession.

Symonides E. (1985a): Changes in phytocoenose structure in early phases of old-field succession in Poland.—Tuexenia 5: 259–271.

Symonides E. (1985b): Floristic richness, diversity, dominance and species evenness in old-field successional ecosystems.—Ekol. Pol. 33: 61–79.

Symonides E. (1986): Seed bank in old-field successional ecosystems.—Ekol. Pol. 34: 3–29.

Symonides E. and Borowiecka M. (1985): Plant biomass structure in old-field successional ecosystems.—Ekol. Pol. 33: 81–102.

Tilman D. (1982): Resource competition and community structure.—Princeton University Press, Princeton, N. J., 296 pp.

Tilman D. (1985): The resource-ratio hypothesis of plant succession.—Amer. Nat. 125: 827–852.

van der Maarel E. (1988): Vegetation dynamics: patterns in time and space.—Vegetatio 77: 7–19.

Vermeer J. G. and Berendse F. (1983): The relationship between nutrient availability, shoot biomass and species richness in grassland and wetland communities.—Vegetatio 53: 121–126.

Wherry E. T. (1920): Soil acidity and a field method for its measurement.—Ecology 1: 160–173.

Whittaker R. H. (1977): Evolution of species diversity in land communities.—Evolutionary Biology 6: 1–67.

Krahulec F., Agnew A.D.Q., Agnew S. & Willems J.H. [eds.]: *Spatial processes in plant communities*. pp. 215–235.

Changes in dry grassland after cutting of Scots pine in inland dunes near Kootwijk, the Netherlands

Fred J.A. Daniels

Institut für Botanik, AG Geobotanik, Schlossgarten 3, 4400 Münster, BRD

Keywords: *Agrostietum coarctatae*, Decorana, Disturbance, Dry grassland, *Festuca tenuifolia*, Inland dunes, Lichens, Mosses, Scots pine, Similarity index, *Spergulo-Corynephoretum cladonietosum*, Succession.

Abstract. Vegetational changes in dry grassland vegetation in inland dunes near Kootwijk, the Netherlands, were annually studied by means of permanent plots during the period 1982/1989. The changes were evaluated using similarity indices and Decorana.

Undisturbed lichen-rich *Spergulo-Corynephoretum cladonietosum* plots appeared to be hardly changed. Successional trends could not be observed. However, cyclic time-related changes could be demonstrated suggesting an internal cyclic dynamic related to the population dynamics of the constituent species.

Agrostietum coarctatae and *Festuca tenuifolia* vegetation disturbed before 1982 by burning, litter dumping and tractors removing cut Scots pines show strong changes, which are linear and time-related. The open *Agrostietum coarctatae* developed into a more closed community with *Festuca tenuifolia* and *Campylopus introflexus*.

Generally a convergent development could be demonstrated; the floristic differences between the grassland communities became less pronounced.

Introduction

The area around the village of Kootwijk, Veluwe, is one of the most extensive eolian drift sand landscapes of the Netherlands and western Europe. The undulating landscape with blow-outs and dunes consists of bare drift-sand tracts and stabilized parts with sparse dry grassland and Scots pine vegetation.

The potential natural vegetation of this atlantic desert area (Schimmel 1975) is *Quercion robori-petraeae* woodland. However human overexploitation of the former oakwoods and heathlands since 1150–1250 A.D. caused the poor, dry pleistocenic cover sands to become exposed to the wind. The final result was a vast desert-like landscape with continuously expanding and redeposited drift sand tracts with sparse vegetation on the more stabilized parts (de Smidt 1969, Koster 1978). However at the end of the last century and the beginning of this one, the extension of the active drift sand area was considerably reduced as a result of Scots pine planting.

Because of its great historical, geomorphological and biological importance a representative part of the area is protected now as a nature reserve. In particular the pioneer association *Spergulo-Corynephoretum* (SC) is optimally developed in all its successional stages and is very rich in lichens. Other plant communities are *Agrostietum coarctatae* (Ac), *Festuca tenuifolia* vegetation (Ft), *Genisto-Callunetum* (GC), Pine forest (Pf) and locally oakwood.

The communities are related to each other in space and time, constituting a successional complex in a dynamic environment.

The existence of the anthropogenic oligotrophic dry grassland vegetation rich in lichens is threatened by succession, which leads, at least temporarily, towards a species-poor pine wood (later *Quercion robori-petraeae* wood). As a pioneer association SC is dependent on the continuous availability of bare sand surfaces. Moreover pine growth must be suppressed or prevented in order to keep the grassland vegetation. As management measures, recreation pressure is allowed to be temporarily intensive in order to create bare sand surfaces, and young pines are regularly removed by hand and adult trees cut.

An additional threat might be the increased level of atmospheric deposition of nutrients, which might directly or indirectly influence the vegetation and succession in these oligotrophic ecosystems. Increase of N has dramatically changed heathland into species-poor grassland in the Netherlands in the last 20 years (cf. Heil 1984, Berdowski 1987). We might also expect influences on this dry grassland vegetation (cf. Daniels *et al.* 1987).

However they are less evident and little is actually known about changes in terms of short term changes, fluctuations and succession, since detailed observations in permanent plots are lacking. Also effects of management, such as removal of pines, burning etc. have hardly been studied.

In order to get more insight in these problems permanent plot observations were started in 1981 (Daniels *et al.* 1987).

Moreover 12 permanent plots were established in 1982 in two inland dunes areas near Kootwijk. In both terrains pine was removed in the second half of the 1970's to favor the lichen rich plant communities.

The present paper will describe and evaluate the annual changes in the 12 plots during the period 1982/1989.

Field sites and permanent plots

The two sites of research are situated north of the village of Kootwijk at the Veluwe, just north of the motor road Amersfoort–Apeldoorn.

Plots CW1/CW6 are situated in the Caitwickerzand (CW). This terrain is a blow-out area with SC, Ac and Ft as dominating vegetation types. Pines were cut here in 1975.

Plots KW1/KW6 are situated in the Kootwijkerzand (KW). This terrain is less vast and more sheltered by surrounding pine forest. It consists of small blow-outs and dunes. *Agrostis vinealis (A. coarctata)* and *Festuca tenuifolia* are more prominent here. Pines were cut in 1979.

In both areas grazing pressure is negligible.

The floristic composition of the 12 permanent plots is shown in Tables 3 to 8.

CW1 and CW3 are Ft stands near pine stumps. Twigs and branches of pines were burnt in the plots in 1975. The lichens are extremely well developed here.

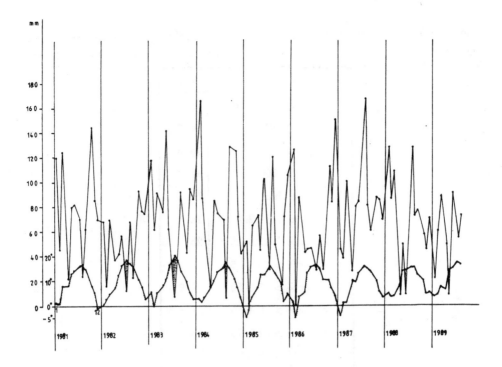

Fig. 1. Mean monthly temperature in °C and mean monthly precipitation in mm over the years 1982-1989. 1: January, 12: December. Weather station Deelen.

CW4 is a *Festuca tenuifolia-Cladina portentosa* vegetation near a pine stump.

CW2, CW5, CW6 and KW2 are SC *cladonietosum* stands in blow-outs (CW2 and CW6) and on low dunes (CW5 and KW2).

KW1 is a bare sand spot in 1982, probably to be classified before 1979 as Ac. In that year soil surface was disturbed by tractors removing cut pines.

KW3 is a *Festuca tenuifolia-Campylopus introflexus* stand on a low stabilized dune with scattered pine stumps of 1979.

KW4 is an Ac stand. Also here soil surface was damaged by tractors in 1979. In 1982 open soil still covered more than 90%. It is situated on a low stabilized dune.

KW5 belongs to Ft and is situated near a pine stump . Twigs, branches and needles were dumped in the vegetation in that year.

KW6 belongs to the GC typicum typical variant (de Smidt 1977) and is situated on the northern slope (15°) of a low (3 m) dune. In 1982 *Calluna vulgaris* was in the building phase (cf. Gimingham 1968).

With the exception of the two fire spots (CW1 and CW3) nutrient content of the soil at −3 cm is very low. P content in 1988 was between 27 and 60 ppm; pH (water) was between 3.8 and 4.3. For the fire spots these values were respectively 155 ppm and more and 6.2 and 7.2. Total N content of the soil under the unburnt sites yielded values around

217

Table 1. (a) abundance/cover scale; (b) Decorana transformation; (c) mean cover % transformation.

	a	b	c
1 or 2 group(s) of shoot(s)	r	1	1
3–10	+	2	2
11–50	1a	3	3
51–100	1b	3	3
more than 100	2m	4	4
cover % 5–12.5	2a	5	9
cover % > 12.5–25	2b	6	19
cover % > 25–37.5	3a	7	31
cover % > 37.5–50	3b	7	44
cover % > 50–62.5	4a	8	56
cover % > 62.5–75	4b	8	69
cover % > 75–87.5	5a	9	81
cover % > 87.5–100	5b	9	94

700 ppm (cf. Daniels *et al.* 1987).

The altitude lies between 20 and 30 m a.s.l.. Ground water is deep and out of reach of the vegetation.

The terrains are situated in the subatlantic climate district (Barkman 1958). The monthly mean temperature and precipitation for the years 1982/1989 in the nearest weather station Deelen, 20 km southeast of Kootwijk, are given in Fig. 1.

Methods

The 12 1 × 1 m plots were recorded every year with the exception of 1988. The records were made in autumn under as far as possible uniform weather conditions. A grid divided into 100 subquadrats was used to estimate more accurately abundance/cover values of the species (See Table 1).

Vegetational changes have been evaluated by means of similarity indices and Decorana computations. Similarity indices between the 1982 and 1989 plot pairs have been calculated after Sörensen (1948), Barkman (1958) and van der Maarel (1966).

Sörensen: $S = (2c/(a + b)) \times 100$; a = total number of species in relevé a, b idem for b and c = number of species shared by relevés a and b.

Barkman: $S = C/\sqrt{(A \times B)}$ and van der Maarel: $S = C^2/(A + C)(B + C)$; A = surplus of values of all species with higher values in relevé a, B idem for b and $C = \sum$ of

218

values common to relevés a and b.

For the calculation of the indices after Barkman and van der Maarel the abundance/cover values have been transformed. One transformation is the same as for the Decorana computations (importance values 1–9), in the other mean cover percentages are used (See Table 1).

Decorana (Hill 1979) was used in the modified version by ter Braak (1982) and van Tongeren (1986) (Standard version, no special options used). Only the first two axes are depicted in the scatter diagrams.

The nomenclature of vascular plants follows Heukels and van der Meyden (1983), lichens Brand *et al.* (1988) and mosses Margadant and During (1982). Nomenclature of syntaxa follows Westhoff and Den Held (1969) and Schröder (1989).

Since it is impossible to judge accurately abundance/cover values of some lichen species separately some are taken together.

Cladonia chlorophaea * includes *C. chlorophaea* s.l., *C. ramulosa*, *C. pyxidata*, *C. verticillata*, *C. strepsilis*; *C. floerkeana* * *C. floerkeana*, *C. glauca*, *C. subulata*, *C. bacillaris*, *C. macilenta*; *C. gracilis* * is mainly *C. crispata* var. *cetrarieformis* and *Coelocaulon muricatum* might include *C. aculeatum*. The occurrence of the algae *Klebshormidium* (B. Pet) Lokh. and *Palmogloea protuberans* (Sm. et Son.) Kütz in the plots was ignored.

Results and Discussion

Remarks on the plots in 1982

The interpretation of the first two Decorana axes (Figs. 2 and 3) in terms of ecological gradients is based on our field data and habitat preferences of the species derived from literature (Ellenberg 1979, Wirth 1980, Touw and Rubers 1989).

Axis 1 is supposed to be a nutrient axis, while axis 2 is supposed to be related to the organic matter content of the uppersoil (which also probably correlates with the moisture conditions). Regarding the high eigenvalues of these two axes (0.6 and 0.35), we assume the floristic variation of the communities on these poor sandy soils to be primarily controlled by nutrient and humus (moisture) content of the uppersoil (cf. also Stoutjesdijk 1959, Lache 1976, Schröder 1989).

A short additional note is made on the lichens. The lichens are extremely well developed in the two fire spots (CW1 and CW3). In particular *Cladonia verticillata* (3 cup levels), *C. subulata* and *C. ramulosa* show high vitality compared with the lichens in the other stands. Moreover most species, especially *C. coccifera*, *C. chlorophaea* and *C. floerkeana* bear abundantly fruiting bodies. It seems that burning has a positive influence on the terricolous "surface" *Cladonia*'s, which develop abundantly seven years after fire (cf. Ahti 1977).

Plot similarity between 1982 and 1989

Similarity indices (See Table 2).

Considering Sörensen's similarity coefficients and taking arbitrarily a similarity of 70 as a boundary, we observe low values for the strongly disturbed plots KW5, KW4 and KW1, but also for KW6, the GC stand. The fire spots (CW1 and CW3) show higher

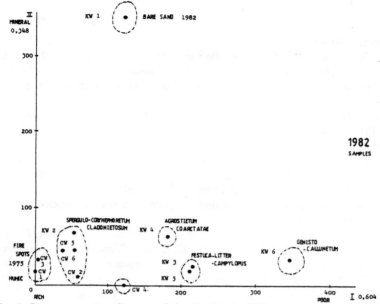

Fig. 2. Decorana scatter diagram of the 12 plots in 1982.

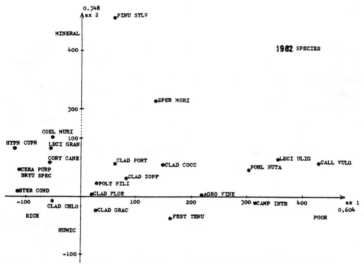

Fig. 3. Decorana scatterdiagram of the species in the 12 plots in 1982.

values, while the undisturbed SC *cladonietosum* stands (CW4, CW5 and CW6) show hardly any differences in floristic composition between the years 1982 and 1989.

However, if we rightly take also into account the quantities of the species in the calculations of similarity, we get another picture. Both Barkman's indices and van der Maarel's indices are low for all plots with former disturbance, including the GC stand in the building phase: CW3, KW6, CW1, KW5, KW4 and KW1. Also the sequence of the plots

Table 2. Similarity coefficients between the plots in 1982 and 1989. (1) Sörensen; (2) Barkman, Decorana transformation; (3) van der Maarel, Decorana transformation; (4) Barkman, mean cover % transformation; (5) van der Maarel, mean cover % transformation. For further explanation see text.

1		2		3		4		5	
CW5	95.2	CW5	12.33	KW2	0.77	CW5	13.10	CW5	0.78
CW4	91.7	CW6	9.40	CW5	0.75	KW2	13.04	CW6	0.77
CW6	91.6	KW2	4.92	CW6	0.66	CW6	10.87	KW2	0.63
KW3	85.7	CW2	4.80	CW2	0.49	CW2	8.36	CW4	0.58
CW3	81.5	CW4	3.21	KW3	0.43	KW3	1.84	KW3	0.57
CW1	80.0	KW3	3.04	CW4	0.30	CW4	1.19	CW2	0.49
KW2	77.8	CW3	2.18	CW3	0.26	CW3	1.03	CW3	0.48
CW2	70.6	KW6	2.00	KW6	0.20	KW6	0.84	KW6	0.40
KW5	66.7	CW1	1.53	CW1	0.12	KW5	0.63	CW1	0.39
KW4	57.1	KW5	1.06	KW5	0.12	CW1	0.53	KW5	0.20
KW6	57.1	KW4	0.94	KW4	0.05	KW4	0.52	KW4	0.14
KW1	00.0	KW1	0.00	KW1	0.00	KW1	0.00	KW1	0.00

according to similarity values in this group is almost the same. The highest similarities are shown by the SC *cladonietosum* stands CW5, CW6 and KW2, while the other stands CW2 (SC *cladonietosum*), CW4 (*Festuca-Cladina* vegetation) and KW3 (*Festuca-Campylopus* vegetation) show intermediate values.

Thus we conclude that the formerly disturbed plots are strongly changed, the undisturbed ones show little change. The *Spergulo-Corynephoretum cladonietosum* stands are hardly changed in seven years.

Decorana diagrams

Since the lengths of the arrows in the scatter diagram (See Fig. 4) inversely reflect the degree of similarity we may conclude that the diagrams confirm the conclusions given above.

Disregarding the outlier KW6, *Genisto-Callunetum*, we also observe the positions of all plots in 1989 situated more in the center of the diagram. Thus a certain convergence among the different grassland stands can be seen. This displacement to the center can be seen very clearly for the disturbed plots, but also the SC *cladonietosum* stands show this feature, although less strongly. This middle part of the diagram could be interpreted ecologically as moderate in terms of humus and nutrient content of the soil. The nutrients (N, P and K) released by the biomass after burning (CW1 and CW3) are gradually leached out in the course of the years; on the bare sand surfaces accumulation of humus takes place due to the increase of vegetation cover (KW1 and KW4); the raw litter decomposes slowly (KW5).

221

Table 3a. Plot CW1 in the years 1982/1989.

CW1: *Festuca tenuifolia* fire spot							
Year	1982	83	84	85	86	87	89
Reference number	1	2	3	4	5	6	7
Total cover %	98	98	100	100	100	100	100
Festuca tenuifolia	2b	2b	2b	3a	3b	3a	4a
Corynephorus canescens	1a	1a	1a	+	+	r	r
Agrostis vinealis	+	r	+	+	r	1a	+
Ceratodon purpureus	4a	2b	2a	1a	1a	+	+
Polytrichum piliferum	2m	3a	4a	3b	3a	3a	4a
Cladonia coccifera	1b	2m	2a	2b	2a	2a	2b
*Cladonia floerkeana**	2a	2a	2a	2a	2b	2b	2b
*Cladonia chlorophaea**	2a	2a	3a	3a	3a	3a	2b
*Cladonia gracilis**	r	+	1a	1a	2a	1a	2a
Cladina portentosa	r	r	r	r	+	+	+
Cladonia zopfii	.	r	r	r	.	r	.
Bryum sp.	2a	.	.	.	.	.	.
Stereocaulon condensatum	r	r	r	r	.	.	.
Rumex acetosella	.	.	.	.	.	+	1a
Cephaloziella sp.	.	.	.	.	.	r	1a
Campylopus introflexus	.	.	.	.	.	.	+

In the undisturbed SC *cladonietosum* stands (CW2, CW5, CW6 and KW2) slow humus accumulation might be expected by the increase of the grasses (See Tables 3, 5, and 6) since 1987. In general we might conclude that almost all stands show a convergent development in the direction of a more grass-rich vegetation. The extreme habitat conditions, nutrient-rich, nutrient-poor, humus-rich and humus-poor seem to be replaced by more intermediate conditions.

Plot dynamics 1982/1989

Relevés (See Tables 3 to 8)

With the exception of the KW2 plot, we see an increase of grasses in the SC *cladonietosum* plots (CW2, CW5, CW6 and KW2). In particular the differences between 1987 and 1989 illustrate this phenomenon. Moreover we observe fluctuations in the abundance/cover values of many cryptogams. *Cladonia chlorophaea* and *C. coccifera* seem to increase. In CW5 and CW6 *Spergula morisonii* as a winter annual might have suffered from the cold winters of 1985/1987, since it is absent in these years or poorly represented.

Table 3b. Plot CW2 in the years 1982/1989.

CW2: *Spergulo-Corynephoretum cladonietosum*							
Year	1982	83	84	85	85	87	89
Reference number	8	9	10	11	12	13	14
Total cover %	50	75	100	100	100	100	100
Festuca tenuifolia	1a	+	+	1a	1a	1a	2a
Polytrichum piliferum	4a	5a	5b	5b	5b	5b	5a
Cladonia coccifera	2m	1b	2m	2m	2a	2a	2b
Cladonia floerkeana*	1a	1a	1a	1a	1a	1a	2a
Cladonia gracilis*	+	1a	1a	1a	1a	1a	1a
Cladonia chlorophaea*	+	+	+	+	1a	1a	2a
Cladina portentosa	r	+	r	r	r	+	.
Corynephorus canescens	.	r	+	+	+	+	+
Spergula morisonii	.	+	1a	1a	1a	1a	1b
Cladonia zopfii	.	r	r	r	.	.	.
Agrostis vinealis	.	.	.	.	.	.	1a
Lecidea uliginosa	.	.	.	.	.	.	+

In the *Festuca-Cladina* plot (CW4) we observe a strong decrease in abundance/cover value for *Festuca tenuifolia*. The dramatic decrease from 4a (1982) to 2a (1989) is very pronounced in 1985/1987, the period with the very cold winters (See Fig. 1). We think this oceanic species (Ellenberg 1979) suffered strongly in these years and died off for the greater part. This phenomenon was commonly observed elsewhere in the Veluwe region. The litter production (of the dead *Festuca*) favoured *Cladonia chlorophaea* as humicolous species, while *Cladina portentosa*, a "volume" lichen decreased in cover value and vitality. The decrease of *Cladina portentosa* in some inland dune sites and heathlands has in our opinion nothing directly to do with atmospheric pollution (cf. a.o. Runge 1984). Changes in vegetation structure including the destruction of the *Festuca* canopy by severe winters without snow, have changed the microclimate and more probably caused the decrease of the aerohygrophytic *Cladina portentosa*. In other stands with a half open, undisturbed canopy of grasses, especially *Agrostis vinealis, Cladina portentosa* thrives extremely well (See also Table 4a, plot CW3).

In the *Festuca-Campylopus introflexus* plot (KW3) a strong decrease of *Campylopus introflexus* is obvious since 1984. As a consequence humicolous cryptogams such as *Cladonia floerkeana* and *Lecidea uliginosa* strongly increased in the cold winter years, growing on the dead *Campylopus*. *Cladonia chlorophaea* and *C. gracilis* also tend to increase.

In the *Genisto-Callunetum* stand (KW6) many changes have occurred. The cover percentage of *Calluna* changed gradually from 3a to 5a (25 to 85%) and its height from 20 to

223

Table 4a. Plot CW3 in the years 1982/1989.

CW3: *Festuca tenuifolia* fire spot							
Year	1982	83	84	85	86	87	89
Reference number	15	16	17	18	19	20	21
Total cover %	100	100	100	100	100	100	100
Festuca tenuifolia	2b	2b	2b	2b	3a	3a	3b
Corynephorus canescens	1a	1a	1a	+	+	+	+
Ceratodon purpureus	4a	3a	3a	2a	2a	+	+
Polytrichum piliferum	+	1a	2a	3a	3a	3a	3b
Cladina portentosa	+	+	1a	1a	1a	1a	2a
Cladonia zopfii	+	+	1a	1a	+	+	+
Cladonia coccifera	2b	2a	3a	2b	2b	2b	2b
*Cladonia chlorophaea**	3a	2b	2b	2b	2b	2b	2b
*Cladonia gracilis**	1a	1a	1a	2a	2a	2a	2a
*Coelocaulon muricatum**	r	r	+	r	+	+	+
*Cladonia floerkeana**	2a	2a	2b	2a	2b	2b	2b
Bryum sp.	2a	1a	1a	+	.	+	.
Agrostis vinealis	+	+	+	r	r	.	.
Hypogymnia physodes	r	.	r	.	.	.	.
Spergula morisonii	+	.	.	.	.	.	.
Cladina arbuscula	.	r	.	.	.	.	.
Campylopus introflexus	.	.	.	r	.	.	.
Lecidea uliginosa	.	.	.	.	+	+	.
Cephaloziella sp.	.	.	.	.	.	.	1a

45 cm. *Campylopus introflexus* decreased, probably due to the unfavorable light conditions under the closed canopy of *Calluna*. As a consequence moisture preferring species such as *Dicranum scoparium* and *Pohlia nutans* appeared, just as did the liverworts *Cephaloziella* and *Gymnocolea inflata*.

In the *Festuca tenuifolia* fire spots (CW1 and CW3) we observe an increase of *Festuca tenuifolia*. Apparently the population did not suffer from the cold winters (cf.CW4) on the relatively nutrient-rich soil (compensation ?). *Corynephorus canescens* however decreased since 1985. The decrease of the eutraphent *Ceratodon purpureus* indicates the expiring of the nutrient pulse after fire in the course of time. In both stands *Polytrichum piliferum* strongly increased, so did *Cladonia gracilis* and *Cladina portentosa* as "volume" lichens (Daniels *et al.*1987), due to the humus accumulation and the more dense *Festuca* canopy

224

Table 4b. Plot CW4 in the years 1982/1989.

CW4: *Festuca tenuifolia-Cladina portentosa* vegetation							
Year	1982	83	84	85	86	87	89
Reference number	22	23	24	25	26	27	28
Total cover%	100	85	100	100	95	100	96
Festuca tenuifolia	4a	3a	3a	2b	2b	2a	2a
Agrostis vinealis	1b	2a	1b	2a	1b	1a	2b
Cladina portentosa	2b	2b	2a	2a	2a	2a	2a
*Cladonia gracilis**	1a	1a	1a	1a	1b	1a	1a
Cladonia coccifera	1b	2a	2a	1a	2a	2a	2b
*Cladonia floerkeana**	2a	2a	2a	2a	2a	2a	2b
Cladonia zopfii	r	r	r	r	r	r	r
*Cladonia chlorophaea**	1a	+	+	1a	2a	1b	+
Polytrichum piliferum	2b	2b	2b	2a	2a	2a	2b
Campylopus introflexus	2a	1a	2a	2a	2a	2a	2b
Spergula morisonii	+	1a	1a	+	.	1a	1a
Corynephorus canescens	r	.	.	.	.	.	.
Cephaloziella sp.	.	1a	.	.	.	.	.
*Coelocaulon muricatum**	.	.	.	.	r	.	.
Lecidea uliginosa	.	.	.	.	.	1a	1a

creating a more air-humid microclimate.

In the *Festuca*-litter plot (KW5) we see a strong increase of *Agrostis vinealis* and *Cladonia coccifera*, while *Spergula morisonii*, *Campylopus introflexus* and *Pohlia nutans* seem to have suffered strongly from the cold winters. The humicolous crustose lichen *Lecidea uliginosa* has high cover values in this cold period favoured by the litter production of *Campylopus* and *Pohlia*. Total cover changed from 35 to 95%, and the raw litter partly decomposed.

The strongest changes are met with in the "open soil" plots KW1 and KW4. In the *Agrostietum coarctatae* (KW4) % cover strongly increased. All grasses, *Festuca, Agrostis* and *Corynephorus* increased, just as did the acrocarpous mosses *Campylopus introflexus, Pohlia nutans* and *Polytrichum piliferum*. The same applies to the humicolous lichens *Cladonia floerkeana* and *C. chlorophaea*. In the bare sand plot KW1 the same increase could be observed. However lichens appeared here in 1987. *Spergula morisonii* seems to have suffered from the cold winters having (almost) disappeared since 1984.

Direction of changes

The Decorana scatter diagrams of the individual plots in all years inform us about

225

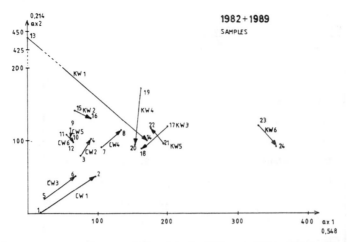

Fig. 4. Decorana scatterdiagram of the plots in 1982 and 1989. The lower number along the arrows corresponds with 1982, the higher with 1989.

time-related vegetational changes in the individual community stands (See Figs. 5 to 8). In the scatter diagrams of the *Spergulo-Corynephoretum cladonietosum* stands (CW2, CW5, CW6 and KW2) we observe a more or less cyclic time-related pattern (Fig. 5). This might suggest a cyclic process in these stands over the seven years, possibly related to the population dynamics of important species. We have seen, that the floristic composition of the stands has changed little if we compare the situation in 1982 and 1989. Thus we might consider the *Spergulo-Corynephoretum cladonietosum* to be a rather stable community, a "Dauergesellschaft", however with an internal cyclic dynamic. However stand KW2 does not fit into this model. The changes are time-related but fluctuations can be observed around the time axis which can be constructed between axes 1 and 2. However the positions of the samples are close together, so also here we see little change. Thus our observations confirm earlier suppositions that the *Spergulo-Corynephoretum cladonietosum* is a floristically stable community (Tüxen 1928, Stoutjesdijk 1959, Schröder 1989).

The fire spots (CW1 and CW3) and bare sand spot (KW1) (See Fig. 6) show a linear time-related pattern along axis 1. This means that the three stands show a linear vegetational change in time and seem to develop into different vegetation types.

The *Genisto-Callunetum* stand (KW6) does not show a clear time related pattern (linear or cyclic) nor do the others (KW3, KW4, KW5 and CW4), whose patterns are capricious (See Figs. 7 and 8).

Succession

Some conclusions on succession might be drawn from the Decorana analysis of all samples 1982/1989 (Fig. 9).

All relevés 1982/1989 of the *Spergulo-Corynephoretum cladonietosum* stands (CW2, CW5 and CW6) still belong to the same cluster, as does the SC *cladonietosum* stand KW2. Thus no succession is observed. The same applies to the fire spots (CW1 and

226

Table 5. Plots CW5 and CW6 in the years 1982/89.

CW5: *Spergulo-Corynephoretum cladonietosum*							
Year	1982	83	84	85	86	87	89
Reference number	29	30	31	32	33	34	35
Total cover %	100	75	90	96	98	95	98
Corynephorus canescens	1a	1a	1a	1a	1a	1a	2a
Festuca tenuifolia	2a	2a	1a	+	1a	+	2a
Polytrichum piliferum	5a	4b	5b	5b	5a	5a	5b
Cladonia coccifera	2m	2m	2b	2b	3a	3a	2b
*Cladonia floerkeana**	1a	1a	2a	2b	2a	2b	2a
*Coelocaulon muricatum**	1a	1a	1a	1a	1b	2a	1a
Cladina portentosa	+	+	+	+	r	+	+
*Cladonia gracilis**	1a	1a	1a	1a	1a	1a	1a
Cladonia zopfii	1a	1a	+	+	+	+	+
Spergula morisonii	1a	1a	1a	.	.	.	1a
Stereocaulon condensatum	.	.	+	r	+	r	.
*Cladonia chlorophaea**	.	.	.	.	.	r	1a
CW6: *Spergulo-Corynephoretum cladonietosum*							
Year	1982	83	84	85	86	87	89
Reference number	36	37	38	39	40	41	42
Total cover %	95	85	95	94	95	100	98
Corynephorus canescens	2a	1a	1a	1a	2a	1a	2b
Festuca tenuifolia	2a	2a	1a	1a	2a	2a	2a
Agrostis vinealis	+	+	1a	1a	r	+	1a
Polytrichum piliferum	4b	5a	5b	5a	4a	4a	5a
*Cladonia floerkeana**	2a	1b	1b	2a	2a	2a	2b
Cladonia coccifera	1b	2a	2a	2a	2b	2b	2b
Cladina portentosa	1b	2a	2a	1b	1a	1b	1a
*Coelocaulon muricatum**	1a	1a	1a	1b	1a	1a	1b
*Cladonia chlorophaea**	1b	1b	1a	2a	2b	2a	2a
Spergula morisonii	1a	1b	1b	.	+	.	1a
*Cladonia gracilis**	+	+	+	1a	.	.	1a
Lecidea granulosa	+	r	.	.	.	.	.
Cladonia zopfii	.	.	r	.	.	.	.

227

Table 6. Plots KW1 and KW2 in the years 1982/89.

KW1: (cf.*Agrostietum coarctatae*; bare sand)							
Year	1982	83	84	85	86	87	89
Reference number	43	44	45	46	47	48	49
Total cover %	4	8	60	75	85	85	98
Pinus sylvestris	r	r	r	r	r	r	.
Spergula morisonii	1a	1b	2m	.	r	.	.
Corynephorus canescens	.	+	2a	2a	2a	+	1a
Festuca tenuifolia	.	+	+	2a	1a	2a	3a
Agrostis vinealis	.	+	2a	3a	3b	3b	2b
Polytrichum piliferum	.	r	+	+	2b	2b	3b
Campylopus introflexus	.	+	+	r	r	+	2a
Pohlia nutans	.	.	+	r	1a	2a	2a
*Cladonia floerkeana**	.	.	.	.	.	+	r
Cladonia coccifera	.	.	.	.	.	.	r
KW2: *Spergulo-Corynephoretum cladonietosum*							
Year	1982	83	84	85	86	87	89
Reference number	50	51	52	53	54	55	56
Total cover %	80	85	88	90	90	90	95
Corynephorus canescens	+	1a	2a	2a	2b	2a	2a
Polytrichum piliferum	5a	5a	5a	5a	5a	5a	5b
*Coelocaulon muricatum**	+	+	1a	+	+	+	r
Cladonia coccifera	2m	1a	2m	1b	1a	1a	1b
*Cladonia floerkeana**	1a	1b	2m	1b	1a	1a	1b
Spergula morisonii	1a	2m	1a	1a	.	1a	1a
Festuca tenuifolia	r	r	r	r	r	.	r
Stereocaulon condensatum	.	r	r	r	r	r	r
Cladonia zopfii	.	.	r	r	r	+	+
Agrostis vinealis	.	.	r	.	.	.	.
Pohlia nutans	.	.	.	r	.	+	+
Cladina portentosa	.	.	.	r	.	.	.
Campylopus introflexus	.	.	.	.	.	.	r

Table 7. Plots KW3 and KW4 in the years 1982/1989.

KW3: *Festuca tenuifolia-Campylopus introflexus* vegetation							
Year	1982	83	84	85	86	87	89
Reference number	57	58	59	60	61	62	63
Total cover %	100	100	98	98	90	98	98
Agrostis vinealis	2a	2m	1b	2a	2a	1a	1a
Festuca tenuifolia	2b	1a	2b	2a	2a	2b	2b
Cladonia coccifera	1a	1a	+	1a	1a	1a	2a
Cladonia zopfii	r	r	+	+	+	+	+
*Cladonia floerkeana**	+	+	+	1b	1b	3a	3b
Lecidea uliginosa	+	+	1a	2b	3b	3a	+
Campylopus introflexus	4b	5a	5b	2b	3a	3b	3b
Polytrichum piliferum	+	1b	1a	+	r	1a	1a
Pohlia nutans	2m	1b	1a	1a	+	1a	1a
Spergula morisonii	+	1a	1a	+	+	1a	.
*Cladonia chlorophaea**	.	.	r	.	r	r	1a
*Cladonia gracilis**	.	.	.	r	r	r	r
KW4: *Agrostietum coarctatae*; open							
Year	1982	83	84	85	86	87	89
Reference number	64	65	66	67	68	69	70
Total cover %	5	10	25	40	75	95	98
Agrostis vinealis	1a	1b	2a	2a	2a	2a	3a
Festuca tenuifolia	r	+	+	+	+	1a	2a
Campylopus introflexus	+	1a	1a	1a	2a	3a	2a
Polytrichum piliferum	+	1a	2a	2b	4a	4b	3b
Spergula morisonii	+	1b	1a	1a	1a	1a	.
Pohlia nutans	.	+	1a	2a	1a	1b	2a
Cladonia coccifera	.	1a	.	1a	1a	1b	1b
Lecidea uliginosa	.	+	1b	1a	2a	2a	.
Corynephorus canescens	.	.	r	1a	1a	1a	2a
Cladina portentosa	.	.	r	.	r	r	.
*Cladonia floerkeana**	.	.	1a	1a	1b	1b	2a
*Cladonia chlorophaea**	.	.	.	r	+	+	+

229

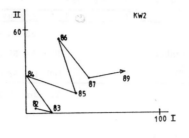

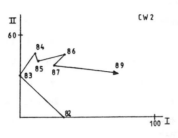

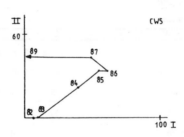

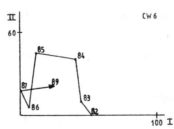

Fig. 5. Decorana scatterdiagrams of the records of each SC cladonietosum plot in the years 1982, 83, 84, 85, 86, 87 and 89.

CW3), the *Festuca-Cladina* stand (CW4), the *Festuca* stands (KW3 and KW5), and the *Genisto-Callunetum* (KW6) stand.

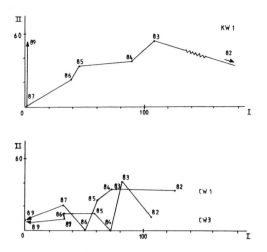

Fig. 6. Decorana scatterdiagrams of the separate records of the plots KW1, CW1 and CW3 in the years 1982, 83, 84, 85 86, 87 and 89.

The relevés of the stands KW1 and KW4 are so close together now, that they might be considered to belong to the same cluster representing one community, a *Agrostietum coarctatae*. Thus succession has taken place in KW1 towards *Agrostietum coarctatae*.

Concluding remarks

In the seven years of observation the undisturbed *Spergulo-Corynephoretum cladonietosum* appears to be rather stable. Cyclic time-related changes occur, but linear successional trends could not be observed. Signs of development into heath or woodland were not observed. In some stands, however, grass cover is fairly high in 1989, but the few relevés do not permit generalisation on an overall increase of grasses. Thus such an overall increase of grasses could not be demonstrated.

With the exception of the *Genisto-Callunetum* (KW6) all other stands show a strong convergent development into *Festuca tenuifolia* dominated communities with *Campylopus introflexus*. Almost all suffered from disturbance before 1982, such as burning, litter dumping and soil damage by tractors. However, *Agrostis* and *Festuca* were already important community components here.

In general an overall increase of *Agrostis vinealis* and *Festuca tenuifolia* might be observed in 1989. Also *Campylopus introflexus* seems to be more common (see also Daniels *et al.* 1987). A negative development of lichens caused by factors other than changes in canopy structure could not be observed in these seven years.

231

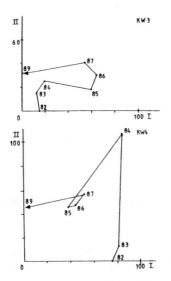

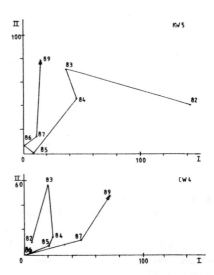

Fig. 7. Decorana scatterdiagrams of the records of the plots KW3, KW4, KW5 and CW4 separately, made in 1982, 83, 84, 85, 86, 87 and 89.

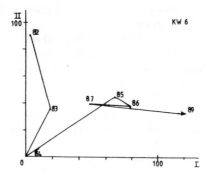

Fig. 8. Decorana scatterdiagram of the records of plot KW6 made in 1982, 83, 84, 85, 86, 87 and 89.

232

Table 8a. Plot KW5 in the years 1982/1989.

KW5: *Festuca tenuifolia* vegetation; litter							
Year	1982	83	84	85	86	87	89
Reference number	71	72	73	74	75	76	77
Total cover %	35	40	50	30	40	70	95
Festuca tenuifolia	3a	1a	2a	2a	2a	2b	2b
Agrostis vinealis	r	+	1a	1b	1a	+	2a
Spergula morisonii	r	1b	1a	r	+	+	1a
Pohlia nutans	2b	1b	2a	+	r	+	1a
*Cladonia gracilis**	r	.	.	+	r	+	+
Lecidea uliginosa	.	1a	2a	2b	3a	3a	1a
*Cladonia floerkeana**	.	1a	1a	1a	2a	2a	3a
Campylopus introflexus	.	3a	2a	1a	r	1a	4a
Cladonia coccifera	.	.	r	1a	1a	1a	1b
Quercus robur	.	.	.	r	.	.	.
Polytrichum piliferum	.	.	.	.	.	.	+

References

Ahti T. (1977): Lichens of the boreal coniferous zone.—In: Seaward.R.D. [ed.], Lichen Ecology, pp. 145-181. Academic Press. London, New York, San Fransisco. 550 pp.

Barkman J.J. (1958): Phytosociology and ecology of cryptogamic epiphytes.—Van Gorcum. Assen. 628 pp.

Berdowski J.J. (1987): The catastrophic death of *Calluna vulgaris* in Dutch heathlands.—Ph.D. Thesis. Elinkwijk. Utrecht. 135 pp.

Brand A.M., Aptroot A., de Bakker A.J., and van Dobben H.F. (1988): Standaardlijst van de nederlandse korstmossen.—Wet. Med. KNNV, nr.188.

Daniels F.J.A., Sloof J.E. and van de Wetering H.T.J. (1987): Veränderungen in der Vegetation der Binnendünen in den Niederlanden.—In: Schubert R. and Hilbig W. [eds.], Erfassung und Bewertung anthropogener Vegetationsveränderungen. Martin Luther Univ. Halle Wittenberg. Wiss. Beiträge, Halle/Saale, 1987/46 (P31): 24-44.

De Smidt J.T. (1969): Het Kootwijkerzand, kerngebied der Europese Stuifzanden.—DLN 72: 7–11.

De Smidt, J.T. (1977): Heathland vegetation in the Netherlands.—Phytocoenologia 4/3: 258–316.

Ellenberg H. (1979): Zeigerwerte der Gefässpflanzen Mitteleuropas.—Scripta Geobotanica IX. 122 pp.

Gimingham C.H. (1968): Ecology of heathlands.—Chapman and Hall. London. 266 pp.

Heil G. (1984): Nutrients and species composition of heathland.—Ph.D. Thesis. Elinkwijk. Utrecht. 139 pp.

233

Table 8b. Plot KW6 in the years 1982/1989.

KW6: *Genisto-Callunetum typicum* typical variant							
Year	1982	83	84	85	86	87	89
Reference number	78	79	80	81	82	83	84
Total cover %	65	98	98	98	100	100	100
Cover % of dwarf shrubs	25	40	60	60	6	80	85
Height of dwarf shrubs cm	20	20	25	30	30	35	45
Calluna vulgaris	3a	3b	4a	4a	4a	5a	5a
Pohlia nutans	+	1a	1a	2m	1a	1b	2a
Campylopus introflexus	3b	5b	5b	5b	5b	4b	2a
Cladonia coccifera	+	r	+	r	+	+	+
Pinus sylvestris	r	+	2a	.	r	r	.
Lecidea uliginosa	+	.	.	.	.	.	.
Cladonia floerkeana*	.	+	r	r	r	.	.
Dicranum scoparium	.	.	+	+	+	1a	2a
Hypnum cupressiforme	.	.	r	.	.	.	.
Festuca tenuifolia	.	.	r	.	r	r	.
Gymnocolea inflata	.	.	.	r	r	.	+
Cephaloziella sp.	.	.	.	r	1a	.	+
Cladonia gracilis*	.	.	.	.	r	.	.
Agrostis vinealis	.	.	.	.	.	.	+

Heukels H. and van der Meyden R. (1983): Flora van Nederland.—Wolters-Noordhoff. Groningen. 582 pp.

Hill M.O. (1979): Decorana. A Fortran program for Detrended Correspondence Analysis and Reciprocal Averaging.—Ithaca, N.Y.Cornell University.

Koster E.A. (1978): The eolian drift sands of the Veluwe (Central Netherlands); a physical geographical study.—Ph.D. Thesis. Univ. of Amsterdam. 195 pp.

Lache D.-W. 1976. Umweltbedingungen von Binnendünen- und Heidegesellschaften im Nordwesten Mitteleuropas.—Scripta Geobotanica XI. 96 pp.

Margadant W.D and During H.J. (1982): Beknopte flora van de Nederlandse Blad- en Levermossen.—Thieme. Zutphen. 517 pp.

Runge F. (1984): Vegetationsschwankungen in einem *Ericetum cladonietosum* II.—Tüxenia 4: 255–256.

Schimmel H. (1975): "Atlantische Woestijnen" de Veluwse zandverstuivingen.—Natuur en Landschap 29: 11–44.

Schröder E. (1989): Der Vegetationskomplex der Sandtrockenrasen in der Westfälischen Bucht.—Abh.Westf. Mus. Naturk. 51/2: 1–94.

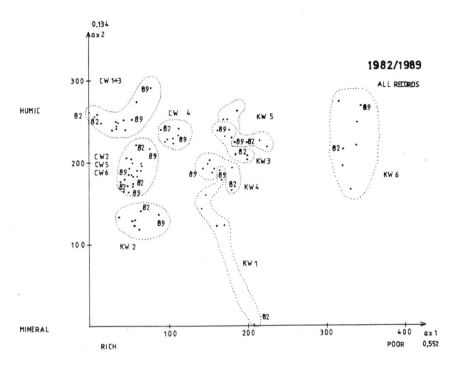

Fig. 9. Decorana scatterdiagram af all records of all plots together of the years 1982, 83, 84, 85, 86, 87 and 89. In the clusters indicated the years 82 and 89 only.

Sørensen T.A. (1948):. A method of establishing groups of equal amplitude in plant sociology based on similarity of species content.—K. Danske Vidensk. Selsk. Biol. Skr.4/4: 1–34.

Stoutjesdijk Ph. (1959): Heaths and inland dunes of the Veluwe.—Ph.D. Thesis. North-Holland Publishing Company. Amsterdam. 96 pp.

Touw A. and Rubers W.V. (1989.): De Nederlandse Bladmossen.—Natuurhist. Bibl. KNNV, Nr.50. Utrecht. 532 pp.

Tüxen R. (1928): Vegetationsstudien im nordwestdeutschen Flachlande. I. Über die Vegetation der nordwestdeutschen Binnendünen.—Jahrb. Geogr. Ges. Hannover 1928: 71–93.

Van der Maarel E. (1966): Dutch studies on coastal sand dune vegetation, especially in the Delta region.—Wentia 15: 47-82.

Westhoff V. and den Held A.J. (1975): Plantengemeenschappen in Nederland.—Thieme. Zutphen. 324 pp.

Wirth V. (1980): Flechtenflora.—UNI-Taschenbücher 1062. Ulmer. Stuttgart. 552 pp.

Krahulec F., Agnew A.D.Q., Agnew S. & Willems J.H. [eds.]: *Spatial processes in plant communities.* pp. 237–249.

Spatial processes in the succession of chalk grassland on old fields in The Netherlands

Jo H. Willems and Roland Bobbink

Department of Plant Ecology and Evolutionary Biology, University of Utrecht, Lange Nieuwstraat 106, NL-3512 PN Utrecht, The Netherlands

Keywords: *Arrhenatheretum elatioris*, Chalk grassland, Digitized maps, Dominance effects, *Mesobrometum erecti*, Old-field succession.

Abstract. The vegetation development on former arable fields in the Wrakelberg Nature Reserve (South-Limburg, The Netherlands) has been monitored by repeated vegetation mapping in the period 1968–85. The management regime, mowing in autumn and removal of the hay, has not been changed during this period.

Four different plant communities could be distinguished all over the period of recording: 1. species rich grassland (*Arrhenatheretum elatioris*) on moderate fertile soil, and 2.–4. three different local types of chalk grassland (*Mesobrometum erecti*).

After digitising the vegetation maps the area of the communities as well as the transitions between the communities in the course of time have been ascertained applying the MAP (Map Analysis Package) computer programm.

The origin and development of the *Arrhenatheretum elatioris* at the lower part of the slope and the *Mesobrometum erecti* on the upper part, strongly depends on the adjacent seed sources. Vegetation succession is dramatically influenced by increasing dominance of a grass species (*Brachypodium pinnatum*) especially during the last observation period (1979–85). Dominance of this grass species is a serious threat for the small scale species diversity. It has been demonstrated that the increasing dominance of *Brachypodium pinnatum* is caused by an external factor, viz. increased nitrogen input by atmospheric pollution since the late seventies.

Based on the digitised maps predictions have been made on the vegetation development in the future.

Introduction

Vegetation mapping, varying from global to microscale, visualises the spatial distribution of vegetation (-complexes), plant species or individual plants. Repeated mapping of the same area with application of similar recording methods can demonstrate successional processes both in space and time. This may enlarge the understanding of ecological processes, predictions for future development may be possible (Kuechler and Zonneveld 1988).

The aim of the present research was to gain insight into the quantitative and qualitative aspects of the succession to a species-rich grassland on old fields adjacent to remnants

237

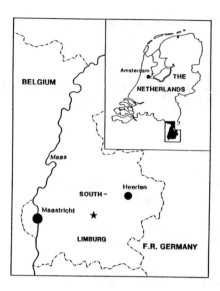

Fig. 1. Location of the Wrakelberg nature reserve in S. Limburg, The Netherlands. Site marked by an asterisk.

of chalk grassland. Knowledge of processes in chalk grassland regeneration on old fields has great importance for nature conservation since the late sixties. From that time onwards areas of marginal agricultural value have been abandoned and often could been acquired by nature conservancy bodies (Bakker 1979). In the Netherlands, much attention has been paid during the recent past to the restoration of species-rich, semi-natural vegetation which was an integrated part of the agricultural system until World War II (Bakker 1987).

In this paper we will describe the development of a chalk grassland on former arable fields, based on four times repeated mapping spread over an 18-year period. Attention will be paid to the pattern of expansion of a dominant grass species caused by external factors. Predictions will be made concerning the future vegetation development in the area.

Study site

The study area, Wrakelberg, is a south-exposed slope (inclination 15–20°) of an east-west running valley, situated in the chalk district in the extreme South of the province of Limburg, The Netherlands (50°52′N; 5°55′E) (Fig. 1). The altitude of the site varies from 130–170m above sea level. The underlying rocks are calcareous deposits of Upper-Senonian (Cretaceous) age (Felder 1974). The soil can be considered as a Rendzina type (De Bakker 1979). Soil depth at the top of the slope is c. 25cm and increases to c. 55cm downhill. The top soil holds c. 4–8% organic matter, 18–33% calcium carbonate and the pH varies from 7–8. The yearly average precipitation is c. 775 mm. The average daily temperatures vary in the coldest month (January) from −1 to +4°C, and in the warmest month (July) from 13 to 23°C. From a structural point of view the vegetation is very patchy and yearly above-ground productivity varies between 150–650 g.m^{-2} dry weight (Verkaar *et al.* 1983a;

238

Bobbink *et al.* 1986).

Of old, the slope has been a common (sheep-)grazing area, as illustrated by maps from the very beginning of the 19th century (Tranchot maps). At the beginning of the present century, probably during World War I, the whole area was ploughed and used as an arable field which was manured regularly, except for a narrow strip at the top of the slope. It is very likely that this part of the site never has been under the plough, because of the undisturbed soil profile. The agricultural crop production (cereals, potatoes, sugarbeets,) ceased in the late fifties (Willems 1987).

The site was set aside as a state-owned Nature Reserve in the autumn of 1961 and from that time onwards managed by the National Forestry Commission. To prevent the site from scrub encroachment and natural succession to woodland the vegetation was mown yearly in autumn (Willems 1983; Bobbink and Willems 1987).

Methods

The vegetation of the site was recorded in 1968, 1973, 1979 and 1985 using the Braun-Blanquet approach in phytosociology (Braun-Blanquet 1964; Barkman *et al.* 1964). The number of vegetation relevés (2 × 2 m) varied from 60 (1973) to 91 (1979). Local vegetation types have been distinguished by tabular comparison of the releves using floristic composition, i.e. differential species.

A grid of permanently marked 20 × 20 m squares was laid out at the site during the monitoring period and the vegetation types were mapped within these squares on a 1 : 100 scale.

All vegetation maps have been digitized as vectormaps by the Analytic Mapping System (AMS) and transformed into gridcell maps (gridcell size 3 × 3 m) by Map Overlay and Statistical System (MOSS). All transitions between the vegetation types in each cell during the monitoring period have been calculated by application of Map Analysis Package (MAP). This program is as AMS and MOSS part of computer program AUTOGIS of the Geography Faculty of the Utrecht University (Burrough and Van der Veer 1984; Ottens and Harts 1985). After the 1985 survey a comparison of all different vegetation types of the previous years was conducted.

Plant nomenclature follows Heukels and Van der Meijden (1983).

Results

A. Vegetation mapping

Four discrete vegetation types which were present during the whole monitoring period have been distinguished in the area:

I. *Arrhenatheretum elatioris* (Class: *Molinio-Arrhenatheretea*)

This type is mainly characterised by grasses like *Arrhenatherum elatius*, *Dactylis glomerata*, *Poa angustifolia*, *Trisetum flavescens*, *Festuca rubra* and the forbs *Senecio erucifolius*, *Tragopogon pratensis*, *Crepis biennis*, *Heracleum sphondylium* and *Knautia arvensis*. At our research site a number of species not typical of this plant association

239

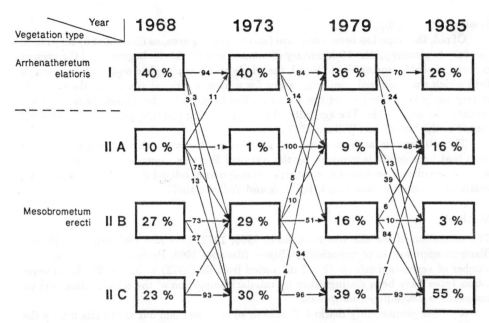

Fig. 2. Diagram showing the area of the site (%) occupied by the four vegetation types at Wrakelberg nature reserve in the different years of vegetation mapping. The arrows indicate the transition percentages of the types during the three intermediate periods. All percentages are based on grid cell map calculations. 2A: intermediate type between *Arrhenatheretum elatioris* and *Mesobrometum erecti*; 2B: type dominated by *Leontodon hispidus* and/or *Briza media*; 2C: *Brachypodium pinnatum* dominated type.

(Westhoff and Den Held 1975), are differential species in this community, viz. *Convolvulus arvensis, Silene vulgaris* and *Potentilla anserina.*

This type is found in the lower part of the slope, where the soil layer is thickest. The *Arrhenatheretum* is well developed on the parts of the slopes where the arable fields occurred up to the late fifties (Fig. 3).

The establishment of this plant community has been favoured by the presence of seed sources in an adjacent species-rich hayfield, belonging to the *Arrhenatheretum elatioris,* which could be found downhill in the early sixties. The mowing regime applied in the nature reserve can be considered as an appropriate management to maintain this association (Ellenberg 1978; Westhoff and Den Held 1975).

II. *Mesobrometum erecti* (Class: *Festuco-Brometea*)

Characteristic of this community are grasses like *Brachypodium pinnatum, Koeleria macrantha* and *K. pyramidata, Avenula pratensis* and forbs like *Scabiosa columbaria, Sanquisorba minor, Gentianella germanica, Thymus pulegioides* and *Ophrys apifera.* All these species can be considered as characteristic species for the association *Mesobrometum erecti* (Willems 1982). Besides, on Wrakelberg a number of species occur frequently in this type,

240

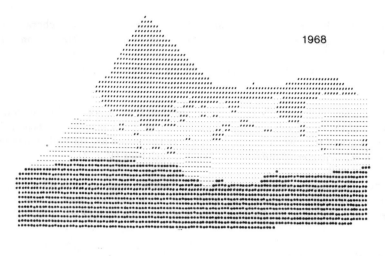

1968

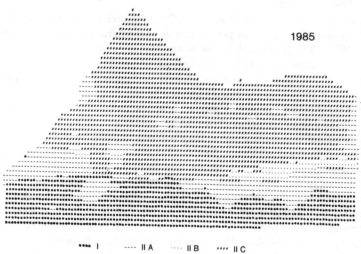

1985

•••• I   ---- II A   ⋯⋯ II B   ⸴⸴⸴ II C

Fig. 3. Grid cell map of the vegetation of Wrakelberg nature reserve in 1968 and 1985 (gridcell size 3 × 3m). I — *Arrhenatheretum elatioris*; II — *Mesobrometum erecti*; A — transitional type between I and II; B — *Leontodon hispidus/Briza media* dominance type; C — *Brachypodium pinnatum* dominated type.

241

which are not characteristic for the association *Mesobrometum*, e.g. *Carex flacca*, *Ononis repens*, *Hieracium pilosella* and *Poa compressa*.

Within the *Mesobrometum* association or chalk grassland community, three types can be distinguished in the Wrakelberg area:

Type IIA. In addition to the *Mesobrometum* species, this type is characterised by number of species frequently found also in the *Arrhenatheretum*, viz. *Tragopogon pratensis*, *Dactylis glomerata*, *Senecio erucifolius*, *Trisetum flavescens* and *Medicago lupulina*. This type can be considered as a transitional one between *Arrhenatheretum* and *Mesobrometum* associations.

Type IIB. This community is clearly distinguished from all others by the dominance (> 50% cover) of the species *Leontodon hispidus* and/or *Briza media*. The type is mainly found on the edges of the former arable fields where the slopes are somewhat steeper than elsewhere as a consequence of different cultivation activities (Fig. 3).

Type IIC. This community is dominated by *Brachypodium pinnatum* (> 50% cover). In the first year of vegetation mapping, 1968, this type was only met at the top of the slope (Figs. 3 and 4). Later on this type spread all over the area (Figs. 4 and 5).

B. Vegetation dynamics

During the vegetation survey of 1968, c. 40% of the area was covered by the *Arrhenatheretum* association (Fig. 3). This percentage decreased gradually in the course of time to c. 26% in 1985. *The Arrhenatheretum* in the Wrakelberg area can be considered as a rather constant community; a high percentage of the area covered by this type showed the same community, especially during the periods 1968–73 and 1973–79, when 94% and 84%, respectively, of their area remained the same vegetation type (Fig. 2). Only during the period 1968–73 was this type supplemented from the intermediate community between *Arrhenatheretum* and *Mesobrometum* (type IIA).

This intermediate community (type IIA) covered c. 10% of the area in 1968 and showed a dramatic fluctuation. From 1968 to 1973, the area covered by this community decreased enormously, whereas in the second and third period this area increased again. The latter was mainly due to a transition of the *Arrhenatheretum* community into this type as a consequence of invasion by species characteristic of types of *Mesobrometum*.

The vegetation dominated by *Leontodon hispidus* and/or *Briza media* (type IIB) showed an obvious decrease from c. 27% of the area in 1968 to c. 3% in 1985. A substantial part of this type originated from the intermediate type (2A) between *Arrhenatheretum* and *Mesobrometum*, especially during the first and third period. Most of this type (2B) has been transformed into the type IIC by invasion of *Brachypodium pinnatum*, especially during the last period of monitoring, viz. 84% (Fig.2).

Brachypodium pinnatum-dominated *Mesobrometum* (type IIC) covered 23% of Wrakelberg area in 1968, and was mainly restricted to the highest part of the slope which has never been ploughed (Figs. 4 and 5). The area covered by this vegetation increased towards c. 55% eighteen years later. Especially during the last period, 1979–85, this expansion was

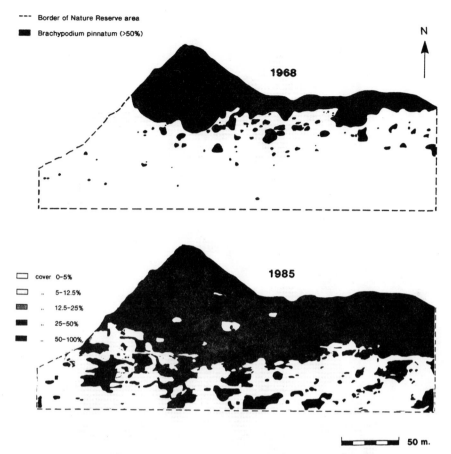

1968

N

1985

50 m.

Fig. 4. Distribution of *Brachypodium pinnatum* at Wrakelberg in 1969 and 1985.

very fast: viz. approximately two times the average percentage in the preceding years. In this period a considerable part (39%) of the intermediate type between *Arrhenatheretum* and *Mesobrometum* (2A) was transformed into the *Brachypodium pinnatum*-dominated vegetation. Where this type became established, it remained very constant: during all periods more than 90% of the occupied area remained the same (Fig. 2).

The invasion of *Brachypodium pinnatum* has taken place in two ways: i. expansion downslope mainly by vegetative spreading of the rhizomes, and ii. colonization far outside the front line by seed germination and established of the seedlings (Figs. 4 and 5).

The average expansion of the front line downhill during the period 1968–73 amounts to 1.73 m yr^{-1} (standard error 0.59). During the next period, 1973–79 this expansion was 1.56m yr^{-1} (s.e. 0.53) and 2.20 m yr^{-1} (s.e. 0.49) during the period before the last mapping in 1985.

The yearly average extension of the area dominated by *Brachypodium pinnatum* ($\geq$

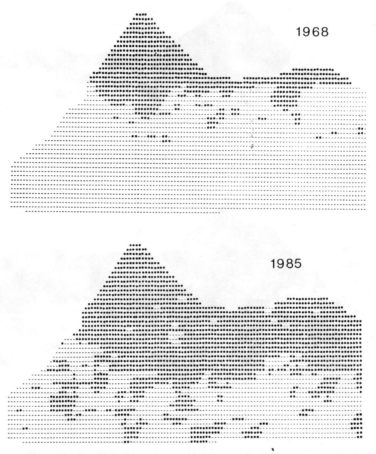

Fig. 5. Grid cell map of *Brachypodium pinnatum* (= cover ≥ 50%) at Wrakelberg in 1968 and 1985 (.).

50% coverage) during the period 1968–73 amounted to 1.73%. For the next period this percentage amounted to 1.95 and 3.17 respectively. These data demonstrate the increased expansion of *Brachypodium pinnatum* during the 18-year period of research. If the latter rate of increase remains the same, almost the whole area of the Wrakelberg (c. 99%) will be covered by *Brachypodium pinnatum* dominated community by the year 2000. The disastrous consequences for the species diversity of the area will be discussed later.

The only spots where *Brachypodium pinnatum* has decreased in abundance or even disappeared completely are patches of scrub scattered over the area (Fig. 4 and 5). This scrub originates from isolated shrubs, mainly *Crataegus monogyna*, *Prunus spinosa* and

Rosa spp., which have been excluded from mowing since the early sixties in order to support the fauna in this area. In the course of time the developing scrub reduced the light quantity near the soil surface and as a consequence the shade-intolerant *Brachypodium pinnatum* (Grime *et al.* 1988) diminished or disappeared under the scrub.

Discussion

Development of species-rich chalk grassland on old fields has been demonstrated to be a rather fast process in the Wrakelberg area. Within a 10-year period after finishing agricultural activities, the area was already covered by highly diverse plant communities belonging to *Mesobrometum erecti* or to *Arrhenatheretum elatioris*. This is not only due to suitable environmental conditions, but also a result of the presence of adjacent seed sources.

The importance of nearby seed sources for vegetation regeneration can be demonstrated by comparing the succession on Wrakelberg with the development at a similar site, Wylre fields, in the same region. The Wylre fields are situated on a N.W. facing slope at a distance of c. 3km from Wrakelberg. The western part of these old fields was abandoned in 1955, whereas on the eastern part of this area cereals have been cropped up to the year 1963. A detailed study of this site has demonstrated that communities belonging to *Mesobrometum* have not developed here, although similarities between Wylre fields and Wrakelberg and another nearby chalk grassland site (Gerendal) at a distance of c. 1.5km, e.g. soil conditions, macroclimate and management regime, are obvious (Hennekens *et al.* 1984). This phenomenon was also noted by Cornish (1954) who found a number of species constantly present all over the district of the North Downs, to be absent in former arable fields aged more than half a century.

Undoubtedly, this results from the absence of adjacent seed sources of *Mesobrometum* species and the very limited seed distribution capacity of many of these species (Verkaar *et al.* 1983b). In 1981 the vegetation of the Wylre fields could be assigned to the *Arrhenatheretum*, although some chalk grassland species were present, too (Hennekens *et al.* 1984). Eight years later this situation has not changed (pers. observ.). It is likely that the vegetation succession of the Wylre fields is determined mainly by the presence of *Arrhenatheretum* species in the adjacent roadside verges and strip lynchets.

Most species of the Wrakelberg spread all over the area during the c. 25 years of conservation management (Fig. 3). This process of fast spreading is undoubtedly a result mainly of the mowing and hay drying activities in the area. However, there are some species which are still met with only in the old strip of never-ploughed chalk grassland at the top of the slope, e.g. *Festuca ovina*, *Carex caryophyllea*, *Koeleria pyramidata*, *Campanula rotundifolia* and *Genista tinctoria*. Seed production in the latter two species could be ascertained almost every year. Probably the germination or establishment conditions of these species are not yet available in the former fields downslope (Senden *et al.* 1986).

A similar observation is mentioned by Cornish (1954) with respect to *Bromus erectus* present in pure stands adjacent to former arable land, but, still absent after a period of 55–70 years of grassland regeneration.

Until the end of the seventies the desired regeneration of chalk grassland on old fields at Wrakelberg was considered to be successful. However, in the early eighties the increase

245

of *Brachypodium pinnatum* in both area and abundance became evident. The expansion in area of *Brachypodium pinnatum* occurred in two different ways: i. spreading of the front downhill, and ii. establishment of new plants outside the frontline. The yearly expansion downhill is not only caused by the speed of spreading of the rizomes, but is also a result of including the *Brachypodium* dominated patches just in front of this line (Fig. 4). This can be based on the considerable variation in the measured spreading distances, as expressed in the standard errors.

The fast expansion and increased abundance of *Brachypodium pinnatum* has been caused by enhanced atmospheric nitrogen deposition (c. 40–50 kg ha^{-1} yr^{-1}) from the late seventies onwards, as a result of industrial and agricultural activities in the region (Bobbink *et al.* 1986, Bobbink 1989).

The dominance of *Brachypodium pinnatum* in the uppermost part of the area (Fig. 4) was not due to an increased nutrient supply, but was a result of the absence of management during the period preceding the early sixties, when the area had been set aside as a nature reserve (Duffey *et al.* 1974; Hakes 1988; Dierschke 1985; Willems 1982, 1985). Later on, this part has been mown in autumn. Autumn mowing, however, is not an effective management strategy in reducing the dominance of *Brachypodium pinnatum*, since in early autumn this grass species redistributes in early auntumn most of its nutrients from the above-ground parts into the extended rhizome system in the soil (Bobbink *et al.* 1989).

In the old grassland strip at the top of the slope, the species diversity was, in spite of the dominance of *Brachypodium pinnatum*, rather high in 1968, viz. 23 species m^{-2}. If *Brachypodium pinnatum* reaches more than 50% of the total dry weight of the vegetation the species diversity declines dramatically (Bobbink and Willems 1987). Although there is some discrepancy between species coverage and dry weight percentage, both values correspond fairly well to each other (Willems 1985). The reason that dominance of *Brachypodium pinnatum* had no dramatic negative effect on the species diversity in 1968 is related to the different growth form of this species at that time, compared with that in 1985. Extra nitrogen supply changes the vertical structure of *Brachypodium pinnatum* as demonstrated in some experiments (Bobbink *et al.* 1988, 1989). The leaves of *Brachypodium pinnatum* are present at a height of more than 0.20 m in a N-fertilised situation and as a consequence overtop most of the species of lower stature. In the unfertilised vegetation dominated by *Brachypodium pinnatum* the leaves are restricted to the lower 0.20 m in the sward. This enables most of the long-lived species to survive (Bobbink *et al.* 1987) and the short-lived species to germinate and establish due to adequate light conditions near the soil surface (Verkaar and Schenkeveld 1984; Senden *et al.* 1986).

The impact on community diversity by a dominant species depends to a high degree on the growth form of the latter (Willems 1985). Therefore, the dominance of the low stature species *Leontodon hispidus* and *Briza media* has no negative effect on the species-richness of the vegetation. The number of species of this vegetation type (IIB) remained the same during the period 1968–85, viz. c. 25 species m^{-2}.

The disastrous impact of the dominance of *Brachypodium pinnatum* on the species diversity of the vegetation first of all can be ascertained on a small scale. The number of species in 0.25 m^2 plots decreased by c. 40% within a 3-year period of nitrogen treatment (Bobbink 1989). However, this effect is not yet reflected in a substantial decrease in the phanerogamic species number in the Wrakelberg area as a whole. This number fluctuated between 140 in 1968 and 132 in 1985. The slight decline is caused by the gradual disappearance of a number of arable weeds during this period.

If the management regime is not modified into mid-summer mowing (Bobbink 1989), the Wrakelberg site will be c. 99% covered by *Brachypodium pinnatum* as the dominant species by the year 2000. This means a serious threat to most of the present species in this nature reserve.

The present study shows that repeated mapping of the vegetation during a long period not only provides information on processes involved in secondary succession, but also can be used in monitoring external impact on the development of an ecosystem. With application of modern mapping and calculation techniques, repeated vegetation recording may help in predicting processes in vegetation development, both in space and time.

Acknowledgements. We would like to express appreciation and gratitude to students in biology at the Utrecht University for their useful and indispensable assistence during fieldwork: Marga Werkhoven (1968), Han Vermeer and Harry Weijs (1973), Marijk den Hoed and Wim Dijkman (1979) and Kristian Kolstrup-Jansen, Wiel Poelmans and Rolf Smeets (1985).

We are grateful to the State Forest Service Department in Limburg province for permission to work in their Wrakelberg area.

References

Bakker P.A. (1979): Vegetation science and nature conservation.—In: Werger M.J.A. [Ed.], The study of vegetation. Dr.W. Junk Publishers, The Hague, Boston, London. pp. 247–288.

Bakker J.P. (1987): Restoration of species-rich grassland after a period of fertilizer application.—In: Van Andel J., Bakker J.P. and Snaydon R.W. [Eds.], Disturbance in grassland. Dr.W. Junk Publishers, The Hague, Boston, London. pp. 185–200.

Barkman J.J., Doing H. and Segal S. (1964): Kritische Bemerkungen und Vorschläge zur quantitativen Vegetationsanalyse.—Acta Bot. Neerl. 13: 394–419.

Bobbink R. (1989): *Brachypodium pinnatum* and species diversity in chalk grassland.—Ph.D. Thesis, University Utrecht.

Bobbink R., Bik L. and Willems J.H. (1988): Effects of nitrogen fertilization on vegetation structure and dominance of *Brachypodium pinnatum* (L.) Beauv. in chalk grassland.—Acta Bot. Neerl. 37: 231–242.

Bobbink R., Den Dubbeldam K. and Willems J.H. (1989): Seasonal dynamics of phytomass and nutrients in chalk grassland.—Oikos 55: 216–224.

Bobbink R., During H.J., Schreurs J., Willems J.H. and Zielman R. (1987): Effects of selective clipping and mowing time on species diversity in chalk grassland.—Folia Geobot. Phytotax. 22: 361–376.

Bobbink R., Van Tooren B.F. and Van Dam D. (1986): Effecten van luchtverontreiniging op kalkgraslandvegetaties.—Natuurhist. Maandbl. 75: 238–242. [with summary in English].

Bobbink R. and Willems J.H. (1987): Increasing dominance of *Brachypodium pinnatum* (L.) Beauv. in chalk grassland: a threat to a species-rich ecosystem.—Biol. Cons. 40: 301–314.

Braun-Blanquet J. (1964): Pflanzensoziologie.—3. Aufl. Springer, Wien, New York.

Burrough P.A. and Van der Veer A.A. (1984): Automated production of landscape maps for physical planning in The Netherlands.—Landscape Planning 11: 205–226.

Cornish M.W. (1954): The origin and structure of the grassland types of the central North Downs.—J. Ecol. 42: 359–374.

De Bakker H. (1979): Major soils and soil regions in the Netherlands.—Dr.W. Junk Publishers, The Hague, Boston, London.

Dierschke H. (1985): Experimentelle Untersuchungen zur Bestandesdynamik von Kalkmagerrasen (*Mesobromion*) in Südniedersachsen, I. Vegetetionsentwicklung auf Dauerflächen 1972–84.— Münsterische Geographische Arbeiten 20: 9–24.

Duffey E., Morris M.G., Sheail J., Ward L.K., Wells D.A. and Wells T.C.E. (1974): Grassland ecology and wildlife management.—Chapman and Hall, London.

Ellenberg H. (1978): Vegetation Mitteleuropas mit den Alpen.—Ulmer, Stuttgart.

Felder W.M. (1974): Lithostratigraphische Gliederung der Oberen Kreide in Sued-Limburg (Niederlande) und den Nachbargebieten.—Publ. Natuurhist. Gen. Limburg 24: 1–43.

Grime J.P., Hodgson J.G. and Hunt R. (1988): Comparative plant ecology. A functional approach to common British species.—Unwin Hyman, London.

Hakes W. (1988): Vergleich de Pflanzenbestandesstruktur genutzer und brachliegender Kalk-Halbtrockenrasen in Nordhessen.— Phytocenologia 16: 289–314.

Hennekens S., Schaminee J. and Westhoff V. (1984): Development of chalk grassland on abandoned fields in South Limburg, The Netherlands.—Colloques Phytosociologiques 11: 471–485. Cramer, Vaduz.

Heukels H. and Van der Meijden R. (1983): Flora van Nederland.—20e Ed. Wolters - Noordhoff, Groningen.

Kuechler A.W. and Zonneveld I.S. (1988): Vegetation mapping.—Handbook for Vegetation Science 10. Kluwer Academic Press, Dordrecht.

Ottens H.F.L. and Harts J.J. (1965): Geografische informatiesystemen en ruimtelijke planning.— Stedenbouw en Volkschuisvesting 25: 26–32.

Senden J.W., Schenkeveld A.J. and Verkaar H.J. (1986): The combined effect and red/far-red ratio on the germination of some short-lived chalk grassland species.—Acta Oecol. 7: 251–259.

Verkaar H.J. and Schenkeveld A.J. (1984): Vegetation structure and the regeneration of short-lived forbs in chalk grassland.—Colloques Phytosociologiques 11: 487–496. Cramer, Vaduz.

Verkaar H.J., Schenkeveld A.J. and Brand J.M. (1983a): Microsite tolerances in relation to vegetation structure.—Vegetatio 52: 91–102.

Verkaar H.J., Schenkeveld A.J. and Van de Klashorst M.P. (1983b): The ecology of short-lived forbs in chalk grasslands: dispersal of seeds.—New Phytol. 95: 335–344.

Westhoff V. and Den Held A.J. (1975): Plantengemeenschappen in Nederland.—Thieme, Zutphen.

Willems J.H. (1982): Phytosociological and geographical survey of *Mesobromion* communities in Western Europe.—Vegetatio 48: 227–240.

Willems J.H. (1983): Species composition and above-ground phytomass in chalk grassland with different management.—Vegetatio 52: 171–180.

Willems J.H. (1985): Growth form spectra and species diversity in permanent grassland plots with different management.—Münsterische Geographische Arbeiten 20: 35–43.

Willems J.H. (1987): Kalkgrasland in Zuid-Limburg.—Wetensch. Meded. 184. K.N.N.V. Hoogwoud. (with English summary).

Krahulec F., Agnew A.D.Q., Agnew S. & Willems J.H. [eds.]: *Spatial processes in plant communities.* pp. 251–259.

Large-scale pattern of biodiversity in Hercynian massifs

Jan Jeník

Institute of Botany, Czechoslovak Academy of Sciences, Dukelská 145, 379 82 Třeboň, Czechoslovakia

Keywords: Anemo-orographic systems, Biogeography, Massif Central, Vosges, Sudeten.

Abstract. A recurring pattern of species-poor and species-rich vegetation in the Hercynian mountains is a result of multiple ecological and palaeo-ecological events induced by topography and wind action. Large air nozzles canalize, accelerate and stir the prevailing western air currents and create a pattern of contrasting factors of climate, soil, hydrology and plant life, i.e., a pattern called 'the anemo-orographic system'. In the lee of funnel-shaped valleys and adjoining treeless summits, a particularly varied mosaic of soil, temperature, moisture and snow enhances differentiation of numerous habitats and plant communities. By preventing the expansion of closed-canopy forests throughout the Postglacial Era, snow avalanches served as a major factor in the survival of many relics and in the establishment of numerous forerunner species.

Introduction

In contrast with the Alps and the Carpathians, the mountains of the Hercynian (Variscan) massifs of Europe (*sensu* Holmes 1965) mostly lack the high altitudes, diversity of rocks and relief complexity. They only exceptionally surpass natural upper limits of forests, and their silicate and base-poor parent rocks seldom create a fertile substratum for eutrophic and calciphilous vegetation. This results in a relatively species-poor oligotrophic flora dominated by competitive grasses, ericaceous shrubs and by a small number of tree species — a feature reinforced by long-term exploitation of forests and subsequent plantation of coniferous monocultures. With regard to this monotonous flora, some of the Central European plant geographers used to write about the 'Hercynian desert' (e.g., Drude 1902).

However the richness and poorness of species vary within any one Hercynian district a great deal. A few districts are enriched by volcanic rocks, but even in silicate ranges, marked centres of higher biodiversity do occur. This can be documented particularly in the large Hercynian massifs which (1) for about two centuries were subject to detailed floristic exploration, collection and taxonomical work, and (2) in recent decades received detailed ecological examination.

The Central Plateau of France (Massif Central), Vosges, Black Forest, Bohemian Forest, Ore Mountains, Sudeten and Harz (Fig. 1) are amongst the best explored mountains

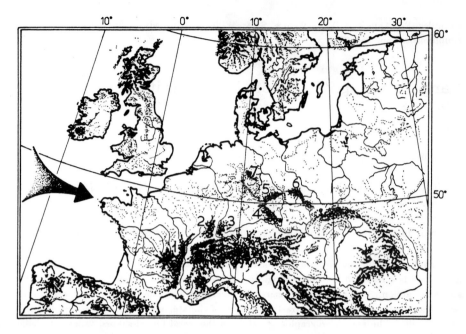

Fig. 1. Situation of the main Hercynian massifs in Western and Central Europe; 1 — Central Plateau, 2 — Vosges, 3 — Black Forest, 4 — Bohemian Forest, 5 — Ore Mountains, 6 — Sudeten, 7 — Harz.

in the world. Older botanical references were summarized by Jeník (1961) and in this paper we merely need to point out the remarkable evidence contributed by recent monographs: Thébaud (1983) and Hennekens *et al.* (1986) from the Central Plateau, Carbiener (1969) and de Valk (1981) from the Vosges, Bogenrieder (1982) from the Black Forest, Sofron and Štěpán (1971) and Petermann and Seibert (1979) from the Bohemian Forest, Heynert (1964) from the Ore Mountains (Krušné hory, Erzgebirge), Šourek (1969) from the Western Sudeten, and Jeník *et al.* (1983) and Bureš *et al.* (1989) from the Eastern Sudeten.

Distributions of plants and mapping of plant communities compiled in the aforementioned monographs suggests that, even within monotonous areas underlain by silicate parent rocks, certain localities harbour exceptional numbers of species, infraspecific taxa and hybrids of vascular plants. Also, from the viewpoint of plant sociology within the large-scale patches of single dominant graminoid, ericoid and coniferous communities, sporadically some herbaceous stands display fine-grained pattern of species-rich vegetation, comparable with that encountered, e.g., in limestone or dolomitic districts.

Due to their proximity to scientific centres and their position astride international boundaries, the Hercynian massifs were subject to multilateral studies, but, so far, only a few of these activities pointed out comparable features in the composition of plant communities (e.g., Carbiener 1969; Oberdorfer 1977, 1978) and similarity in vegetational pattern

252

(Jeník 1961). Remarkable centres of biodiversity in individual Hercynian ranges are treated mostly as random features resulting from random distribution of the underlying bedrocks and local soil moisture.

The anemo-orographic systems of the Sudeten

Thirty years ago, the present author (Jeník 1959, 1961) published a study summarizing ecological and phytosociological observations in the Sudeten. Situated along the boundary between Poland and Czechoslovakia this massif consists of a set of ranges, with three dominant ridges surpassing the present-day upper forest limit (alpine timberline); the high plateaux of the Giant Mts., Králický Sněžník range and Hrubý Jeseník range are at about 1400 m a.s.l.; Sněžka Peak reaches 1603 m. The bedrocks are prevailingly gneiss, phyllite and granite; soil is mostly nutrient-poor and acid. The Sudeten are exposed to western oceanic air currents, and are thus supplied by high totals of precipitation, getting in winter deep snowpack, with occasional avalanches.

The forest flora of the montane belt in the Sudeten displays only little diversity, with European beech (*Fagus sylvatica*) and Norway spruce (*Picea abies* (Sudeten) predominating in the forests. Swiss dwarf pine (*Pinus mugo*) creates the single-dominant krummholz in the subalpine belt only in the Giant Mts., while most summits protruding above the timberline are covered by species-poor grassland dominated by *Calamagrostis villosa*, *Nardus stricta*, *Festuca airoides*, etc., and/or by ericaceous dwarf scrub composed of *Vaccinium myrtillus*, *V. vitis-idaea*, *Calluna vulgaris*, etc. In terms of phytosociological units, the Sudetic summits are mostly covered by comparatively monotonous classes of the *Vaccinio-Piceetea* and *Nardo-Callunetea*.

In many floristic and faunistic works from the Sudeten, however, certain localities are repeatedly referred to, which excel in numbers of plant and animal species. Their remarkable biodiversity consist of a mixture of taxa which are associated both with uplands and lowlands, belong to various centres of origin, represent a variety of phytosociological syntaxa and reflect contrasting habitat and ecosystemic affinities. Moreover, these localities harbour many endemic or subendemic populations, whose systematics remain a permanent challenge to taxonomists.

Following a preliminary study (Jeník 1961) we have undertaken a detailed analysis of some of these centres of diversity in the Giant Mts. (Pančava Corrie, Úpa Corrie, Devil's Garden), and in the Hrubý Jeseník range (Velká Kotlina Cirque). Beside the census of plant species and plant communities, we have observed essential environmental factors, including the snowpack and avalanches in winter (see, e.g., Jeník *et al.* 1980, 1983).

In conclusion we have identified some common features in the position of the species-rich sites. Most of them are (1) situated on east facing slopes, on the eastern margin of upland plateaux, or on the eastern side of prominent saddles; (2) lying across the summit on the opposite side of a funnel-shaped valley; (3) situated in a hollow landform, corrie or cirque; (4) coincident with the position of snow cornices, snow drifts and avalanche tracks in winter.

Prevailing westerly winds stabilized and accelerated air currents in the funnel-shaped valleys and over the corresponding summit or saddle areas (documented by flag-trees, eroded soil and snow ablation forms), and turbulent air in the sheltered lee (documented

Fig. 2. View from the west of the windward funnel-shaped valley and acceleration summit zone of the Bílé Labe A-O System in the Giant Mountains; the arrows mark the upper edge of three leeward turbulence zones serving as centres of biodiversity, from the left the Cirque of Small and Great Lake, Cirque below Sně žka Peak, and Úpa Cirque.

by eddies of fog, clouds and snow drifts) brought us to the description of particular systems of interfering topography and airflow; we have called them 'anemo-orographic systems' (hereafter A-O system). There are about 10 prominent A-O systems in the Sudeten. Each of them consists of three interdependent portions: (1) windward funnel-shaped valley, (2) acceleration summit zone, and (3) leeward turbulence zone (Fig. 2). The A-O system follows an aerodynamic flow similar to that described around small-scale obstacles (Oke 1978), but on a larger scale.

The leeward turbulence zone coincides with the Sudetic centres of biodiversity. Only secondarily can the richness of habitats be reinforced by outcropping base-rich rock, which results from the accelerated water and avalanche erosion. Current richness of flora and fauna is further enhanced by the long-term absence of competitive forests which were continually destroyed by avalanches, even in the climatic optimum of the Postglacial Era. Presence of abundant endemic microspecies and hybrids led us to the hypothesis of 'evolutionary workshops' in the lee of A-O systems (Jeník 1983).

Comparable phenomena in other Hercynian massifs

In a concluding chapter of my monograph on Sudetic A-O systems (Jeník 1961: 342-353) I pointed out some comparable vegetational phenomena in neighbouring mountain ranges. Using first-hand experience and many published accounts it is possible name particular localities in the Vosges, Black Forest, Bohemian Forest and Ore Mountains, where joint

254

effects of topography and local winds suggest a plausible explanation of local biodiversities. After 30 years the evidence for multilaterally functioning A-O systems has increased.

In a number of research papers Carbiener (1969, 1979, 1980) analyzed the *Calamagrostion arundinaceae*, the species-rich and 'subthermophilous' alliance of plant communities encountered in the summit area of the Vosges and other Hercynian massifs. All these tall-grass stands develop in the leeward shelter of large cirques situated on eastern slopes of the main range; an enormous diversity and isolation of the populations resulted in the development of specialized 'ecotypes' of grassland species. De Valk (1981) confirmed that the species-rich tall-forb and tall-grass communities developed in the summit area of the Vosges, uninterrupted by progress of forests in the mid-Holocene age.

In the summit area of the Black Forest evidence for the development of well-known centres of biodiversity also increased. Bogenrieder *et al.* (1982) described a 'subalpine island' in the area of Feldberg from all scientific aspects. Apparently the most species-rich locality, the cirque of the Feldberg Sea, is situated right in the lee of two interfering A-O systems (Brugga A-O System, Wiese A-O System). As in the Sudetic ranges, the biodiversity is not connected with the uppermost altitude of the Feldberg Peak.

Carbiener (1969) was first to point out the affinities of the highly complex grassland associations in the Central Plateau in France, and the other Hercynian massifs. Presenting vegetation relevés from the volcanic Monts Dore and from Monts Domes he supported the hypothesis of similar environmental background and Postglacial development of vegetation in favourably sheltered habitats and refuges. The activities of Dutch botanists in the Monts du Forez at the eastern margin of the Central Plateau (e.g., Coolen and Siebum 1988; Jansen and de Leeuw 1988) produced remarkable evidence of relationships between species-rich vegetation and snow-rich localities, such as the famous Corniche en Coeur and La Grande Courbe.

With regard to the Bohemian Forest Sofron and Štěpán (1971) undertook a detailed examination of the Jezerní Stěna, a locality designated by many botanists as the most interesting area of the whole range. They have confirmed the efficiency of the Weisser Regen A-O System, in creating abnormal microclimatic and soil conditions over the slopes of the cirque, thus producing occasional snow gliding and avalanches, and functioning in the Postglacial florogenesis.

Anemo-orographic system in the Ore Mountains

In the above quoted monograph (Jeník 1961: 342-344) I have suggested the possibility of a functioning A-O system in the Ore Mountains, a relatively low Hercynian range lying astride the Czechoslovak and East Germany boundary. Reaching on the Klínovec Peak a maximum altitude of only 1244m, this range does not surpass the level of the natural timberline, and its major botanical specialities consist merely of montane bogs. It was however Drude (1902: 574, 576) and more recently Siegel (1962) who mentioned the puzzling occurrence of subalpine forb and scrub communities in the valley-head called Zechengrund, at an elevation of only 1050 m.

Topogaphy and wind action are efficient factors in all sizes of landforms, but their effects on plant life are always buffered in those ranges which presently are, or in the past were, covered by a closed-canopy forest. Single dominant stands of Norway spruce (*Picea*

255

abies), particularly, do not leave much space for less competitive heliophilous alpine and subalpine forbs, such as *Epilobium alpestre* (Jacq.) Krocker, *Cicerbita alpina* (L.) Wallr. and *Ranunculus platanifolius* L.

Heynert (1964) explored the timberline and found evidence of flag-trees on the German territory of the Ore Mountains, namely on Fichtelberg, its second highest elevation, and in the area of Zechengrund. Heynert even confirmed my earlier assumption of accelerated airflow through the exposed saddle above the town of Boží Dar, between the elevations of the Fichtelberg and Klínovec.

Following our exploration on both sides of the Czechoslovak-German boundary, and taking into account flag-trees, distribution of snow and occurrence of rare plant populations amidst 'Hercynian desert', we were able recently to define a clear-cut A-O system causing the enhanced biodiversity in the Zechengrund. Taking the name of to the river streaming through the windward funnel-shaped valley, we call this topography/wind pattern the Černá A-O System.

On the topographic map (Fig. 3) we have marked major contourlines, elevations and air currents detected mainly from flag-trees, movements of clouds and distribution of snow. The windward valley of Černá extends in the ideal West-East direction, which enables the system to collect winds of the whole western quadrant. Two pairs of summits, one above 1000 m, the other above 1200 m altitude, converge, squeeze and accelerate the streamlines; downwind, in the lee of this air current, decelerated flow and complex eddies enhance sedimentation of soil, plant seeds and snow, right in Zechengrund, the species-rich locality. Disturbance of the closed-canopy forest and depression of the natural timberline occurred in the past both on the wind-exposed terrain, and in the zone of snow drift.

The summit plateau with the small town of Boží Dar is the windiest and most snow-rich area of this whole mountain area (Fig. 4). The cross-section shows also the moderate gradient surface between about 800 and 1080 m elevation, the latter being on the top of a saddle between the two highest peaks; this kind of slope is ideal for the acceleration of the airflow and for its contrasting ecological action in the windward and leeward zones. Remarkably, the boundary between these zones is a national boundary, and in spite of the declared nature reserve in Zechengrund, a shelterbelt has recently been planted to 'protect' the area against snow drifts.

Some general remarks

Pigot and Walters (1954) when discussing discontinuous distribution of rare plants in Britain stressed that any interpretation must contain both an 'historical' and an 'ecological' element. .This applies generally to mountain vegetation, and to the phenomenon of species-rich localities in the Hercynian massifs, too.

An 'historical' account must involve the time scale of the whole Postglacial Era, and refer to (1) physical environment, including relief changes, soil forming processes and climatic alteration, and (2) biotic phenomena, including microevolution, migration routes, etc. An 'ecological' account should involve various scales which govern diverse biotic and ecosystemic processes.

At this point we need to return to the concept of A-O systems. Billings (1979) described similar orderly pattern in high mountains in North America. Accepting prevailing

256

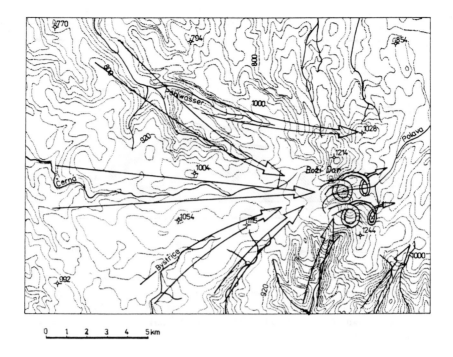

Fig. 3. Topographic sketch-map of the upper summits in the Ore Mountains showing the situation of the Černá A-O System; the arrows mark convergent airflows affecting the acceleration summit zone, and the turbulences generated above the Zechengrund, a centre of snow drifts and plant species diversity.

winds as essential environmental factor, this author has proposed the concept of 'ᴍesoto-pographic units', representing an environmental gradient composed of a windward slope, ridgecrest, lee slope (with snow drift) and meltwater meadow. On a diagram (op.c.: 104) the cross-section of the ridge shows equal surface, inclination and landform on both sides of the ridge. This would suggest (1) that neither vegetation cover nor the nivation process performed a creative role in the pattern, or (2) a very temporary and short existence of the pattern.

Neither of these assumptions seems to be valid for the pattern of diversity studied in the Hercynian mountains. The leeward section of an A-O system here is always markedly sculptured into a more or less deep hollow landform, a result of nivation (started and recurring by late snow patches at any period of a snow-rich climate) and a consequence of glacial erosion performed during the peak of glacial periods; a kind of nivation niche or glacial cirque is necessarily situated in the lee of all A-O systems.

257

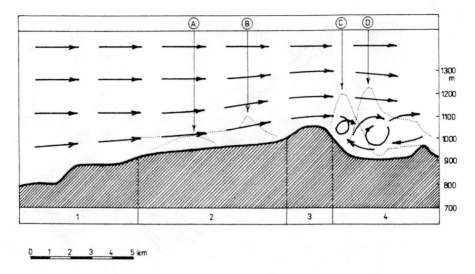

Fig. 4. Cross-section of the Černá A-O System showing the prevailing streamlines of the western wind; 1 — basal zone near the town of Johann-Georgenstadt; 2 — windward funnel-shaped valley; 3 — acceleration summit zone; 4 — leeward turbulence zone over Zechengrund and the valley head of Polava Brook; A, B, and C, D mark two pairs of summits affecting the squeezing of the airflow.

If a satisfactory time-scale is taken into account, the pattern of mountain vegetation should include long-term processes of succession, and even microevolution and natural selection. Carbiener (1979, 1980) provided us with many examples from the Vosges, and biosystematics students continue to describe new small taxa, such as *Sorbus sudetica* (Tausch) Fritsch, *Plantago atrata* subsp. *sudetica* (Pilger) Holub, *Dianthus carthusianorum* L. subsp. *sudeticus* Kovanda, and *Knautia arvensis* (L.) Coulter subsp. *pseudolongifolia* (Szabó) O. Schwarz, whose cradle was in the leeward turbulence zone of the Sudetic A-O systems.

For many years avalanches and snowdrifts are regarded as danger to plant life in mountains. As pointed above the same factor should be considered to have beneficial effects for the rare plant populations in many species-rich localities of Hercynian massifs.

References

Billings W.D. (1979): High mountain ecosystems.—In: Webber P.J.[ed.], High altitude geoecology. Amer. Ass. Advancement of Science, Selected Symposium No. 12.

Bogenrieder A. *et al.* (1982): Der Feldberg im Schwarzwald — subalpine Insel im Mittelgebirge.— Karlsruhe, Landesanstalt Umweltschutz Baden-Württenberg.

Bureš L., Z. Burešová, V. Novák (1989): Vzácné a ohrožené rostliny Jeseníků.—Bruntál, Český Svaz Ochr. Přírody.

Carbiener R. (1969): Subalpine primäre Hochgrassprärien im herzynischen Gebirgsraum

Europas...—Mitt. Flor.-soziol. Arbeitsgemeinschaft, series nova, 14: 322–345.

Carbiener R. (1979, 1980): Étude des écotypes d'espèces collectives praticoles infeodées aux prairies subalpines primaires (Calamagrostion arundinaceae)...—Documents Phytosoc., series nova, 4: 1065–1079, 5: 375–408.

Coolen C. and Siebum M. (1988): Een onderzoek naar de vegetatie van sneeuwnissen in de Monts du Forez...—Nijmegen, Afdeling Exper. Plantenoecologie.

De Valk E.J.(1981): Late Holocene and present vegetation of the Kastelberg (Vosges, France).—Utrecht, Laboratory of Palaeobotany and Palynology.

Drude O. (1902): Der herzynische Florenbezirk.—In: Die Vegetation der Erde, vol.6. Leipzig, W.Engelmann.

Hennekens S. et al. (1986): De vegetatie van de Hautes Chaumes van de Monts du Forez (Frankrijk).—Utrecht, Laboratorium Paleobotanie en Palynologie.

Heynert H. (1964): Das Pflanzenleben des Hohen Westerzgebirges.—Dresden, Verlag Theodor Steinkopff.

Holmes A. (1965): Principles of physical geology.—London, Thomas Nelson and Sons.

Jansen J. and de Leeuw J.P.M. (1988): Een vegetatiekundig onderzoek naar de Sorbus-struwelen in de subalpiene zone van de Monts du Forez (Frankrijk).—Nijmegen, Laboratorium Exper. Plantenoecologie.

Jeník J. (1959): Kurzgefasste Übersicht der Theorie der anemo-orographischen Systeme.—Preslia 31: 337–357.

Jeník J. (1961): Alpinská vegetace Krkonoš, Králického Sněžníku a Hrubého Jeseníku.—Praha, Nakladatelství Čs. Akad.věd.

Jeník J. (1983): Evoluční jeviště sudetských karů.—Biol. Listy 48: 241–248.

Jeník J., Bureš L. and Burešová Z. (1980): Syntaxonomic study of vegetation in Velká Kotlina cirque, the Sudeten Mountains.—Folia Geobot. Phytotax. 15: 1–28.

Jeník J., Bureš L. and Burešová Z. (1983): Revised flora of Velká Kotlina cirque, the Sudeten Mountains, pars 1 et 2.— Preslia 55: 25–61, 123–141.

Oberdorfer E. (1977, 1978)[ed.]: Süddeutsche Pflanzengesellschaften, pars 1 et 2.—Jena, VEB Gustav Fischer.

Oke T.R. (1978): Boundary layer climates, 2nd edition.—London, Methuen and Co. Ltd.

Petermann R., Seibert P. (1979): Die Pflanzengesellschaften des Nationalparks Bayerischer Wald...—Grafenau, Bayerisches Staatsministerium f. Ernährung, Landwirtschaft u.Forsten.

Pigott C.D. and Walters S.M. (1954): On the interpretation of the discontinuous distribution shown by certain British species of open habitats.—J. Ecology 42: 95–116.

Siegel M. (1962): Beiträge zur Ökologie von Pflanzengesellschaften mit Vertretern des arktisch-alpinen Florenelementes im Fichtelberggebiet (oberstes Erzgebirge).—Ber. Arbeitsgem. Sächs. Bot., series nova: 187–216.

Sofron J. and Štěpán J. (1971): Vegetace šumavských karů.—Rozpravy Čs. Akad. Věd, řada mat.-přír. věd, 81/1: 1–57.

Šourek J. (1969): Květena Krkonoš — Český a polský Krkonošský národní park.—Praha, Academia.

Thébaud G. (1983): Contribution á l' étude des hautes-chaumes du Forez: phyto-écologie des communaux du Brugeron (Puy-de-Dome)...—Clermont-Ferrand, l' Université.